Y0-BQR-107

Instructor's Guide for

Introduction to Technical Mathematics

Fourth Edition

Allyn J. Washington
Mario F. Triola

Dutchess Community College
Poughkeepsie, New York

The Benjamin/Cummings Publishing Company, Inc.
Menlo Park, California • Reading, Massachusetts • Don Mills, Ontario
Wokingham, U.K. • Amsterdam • Sydney • Singapore
Tokyo • Madrid • Bogota • Santiago • San Juan

Copyright © 1988 by The Benjamin/Cummings Publishing Company, Inc.

All rights reserved. No part of this publication may be reproduced, stored in a retrieval system, or transmitted, in any form or by any means, electronic, mechanical, photocopying, recording, or otherwise, without the prior written permission of the publisher. Printed in the United States of America. Published simultaneously in Canada.

ISBN 0-8053-9539-3

4 5 6 7 8 9 10 CRC 98979695

The Benjamin/Cummings Publishing Company, Inc.
2727 Sand Hill Road
Menlo Park, California 94025

CONTENTS

ANSWERS TO EXERCISES

PREFACE

This instructor's guide for <u>Introduction to Technical Mathematics</u>, 4th edition, contains general comments and suggestions on the use of the material in each chapter and the answers to the exercises. It is hoped that this information will aid the instructor in preparing and designing the course to fit the needs of the students.

The comments and suggestions include the purpose of each chapter as well as the points which should be emphasized. Difficult topics are identified. It is hoped that this information will alert instructors to student questions and to ideas that may need additional clarification.

One suggestion appropriate to all chapters relates to the sets of review exercises. These are included at the end of each chapter to give the student the opportunity to review frequently. A full class period should be given to each of these sections so that the basic operations may be mastered. Also, the student should be advised to take full advantage of the opportunity to review.

Some comments concern possible organization of the material. Sections that can be omitted without loss of continuity are identified, as are those sections or topics that may require extra class time. This information will help in organizing courses to fit particular needs.

Included on pages 32-33 is a table summarizing the topics covered in Introduction to Technical Mathematics, 4th edition, and Basic Technical Mathematics, 5th edition. This table is included to help the instructor select the text and topics suited to a specific course.

The textbook contains answers to odd-numbered exercises only, but this guide contains answers to all exercises. Some instructors prefer to assign both even-numbered exercises and odd-numbered exercises so they can check student's work. Assigning exercises that are multiples of three, such as 3, 6, 9, etc., will generally provide a good mix of problems that will include both even-numbered and odd-numbered types. Some instructors may wish to have their students obtain this guide so that answers for all exercises will be available for checking.

The text is intended for courses up to two semesters or three quarters of three or four hours of each class week. This estimate is based on covering about one section (including the review exercises) per lesson and on having four or five hours for tests per semester. For shorter courses, it is possible to combine some sections and omit others (as noted in the first discussion in this guide).

We hope that this guide is helpful in obtaining maximum benefit from the text. We strongly encourage you to share with us any suggestions you may have.

COMMENTS AND SUGGESTIONS

Chapter 1   ARITHMETIC

The general purpose of this chapter is to review the basic arithmetic operations of addition, subtraction, multiplication, and division. These operations are reviewed on the whole numbers, fractions, and decimals. The operations introduced in this chapter will be used as a basis for developing later material in algebra.

Section 1-1 introduces the operations of addition and subtraction of whole numbers. To help the student understand why these operations are performed as they are, the concept of positional notation, carrying, and borrowing are discussed. However, the main point to be emphasized to the student is that the basic sums should be known without hesitation.

Multiplication and division of whole numbers are developed in Section 1-2. The associative and commutative laws are mentioned, but emphasis is on the distributive law which shows the reasoning of the process of multiplication.

Fractions are introduced in Section 1-3. Here proper fractions, improper fractions, and mixed numbers are discussed.

Equivalent fractions are discussed in Section 1-4. The main point in this section is that of reducing a fraction to its lowest terms by dividing both numerator and denominator by the same number. This is done directly for simpler numerators and denominators, and the process of factoring the numerator and denominator is covered for use with larger numerators and denominators. If this process is well understood, cancellation in algebraic fractions will be better understood when it is encountered.

In Section 1-5, addition and subtraction of fractions are discussed. Determination of the lowest common denominator should be emphasized. If necessary, additional class time might be devoted to this section. Students are often unable to handle the basic operations on fractions. An understanding of arithmetic fractions is essential to the proper manipulation of algebraic fractions.

Multiplication and division of fractions are considered in Section 1-6. Normally, students do not have as much difficulty with these operations as they do with addition and subtraction of fractions. However, it is important to emphasize the removal of common factors before actually performing the multiplication.

The basic operations on decimals are taken up in Section 1-7. Performing these operations allows the student to continue to develop arithmetic skills.

Percent is the topic of Section 1-8. Again, the operations in this section allow the student to further develop

arithmetic skills. The only difficult procedure is determining the base when the percentage and the rate are given,
as in Example H.

Powers and roots of numbers are introduced in Section
1-9. The basic concepts and notation are important for
later work in algebra.

In Section 1-10, the evaluation of square roots is discussed. The use of calculators is discussed along with the
use of a table. If a calculator is to be the principal
method, then this section may be omitted.

Section 1-11 deals with the use of calculators for the
fundamental arithmetic operations discussed in the pre-
ceding sections of Chapter 1. This section serves as an
important foundation in using the calculator which has
become so necessary for technicians and scientists.

Chapter 2   UNITS OF MEASUREMENT AND APPROXIMATE NUMBERS
One of the purposes of this chapter is to introduce the use
of units, including those in the SI metric system, and unit
conversions, which are useful in allied technical courses.
Also, the student will become accustomed to handling approximate numbers and rounding off results properly. A
beneficial side effect is that the student will continue to
develop arithmetic skills while making conversions and
dealing with approximate numbers.

3

Section 2-1 introduces units of measurement, with emphasis on the SI metric system. The student should become familiar with metric units so that the symbols and prefixes are recognized in later chapters.

Conversion and reduction of units of measurement are discussed in Section 2-2. Many students have difficulty in making unit conversions. The method of conversion illustrated in Examples E, F, and G should be emphasized.

Approximate numbers and significant digits are discussed in Section 2-3. Determination of the significant zeros is the most troublesome point in this section.

In Section 2-4 arithmetic operations on approximate numbers are discussed. It must be emphasized that there is no need or value in keeping extra significant digits when dealing with approximate numbers. For a shorter course, it is possible to combine this section with Section 2-3 for a single lesson.

Chapter 3   SIGNED NUMBERS

This chapter introduces signed numbers and shows the student how to perform the basic operations with them. This skill is necessary for understanding the algebra in later chapters.

Section 3-1 introduces the concept of a signed number and its interpretation. Although subtraction has been previously encountered, it has not been necessary to use negative numbers. Also, the concepts of <u>greater than</u>, <u>less than</u>, and <u>absolute value</u> are introduced.

The addition and subtraction of signed numbers are discussed in Section 3-2. Here it should be emphasized that these operations are generally converted to operations on unsigned numbers, which are then combined.

Multiplication of signed numbers is covered in Section 3-3. The points to emphasize here are that changing the sign of one factor changes the sign of the product, and that sign can be determined by counting the number of negative factors.

Section 3-4 deals with the division of signed numbers. The operations associated with zero can cause students some difficulty. It should be emphasized that only division <u>by zero</u> is undefined; other operations with zero are defined and follow the appropriate rules.

Section 3-5 considers the order of execution of the operations of addition, subtraction, multiplication, division, and powers. As students become more involved with algebra and with calculators and computers, this order of operations becomes more important.

Chapter 4   INTRODUCTION TO ALGEBRA

The material in this chapter provides an introduction to

basic algebraic notation and operations, which will be used and expanded in later chapters.

In Section 4-1 the idea of using literal symbols to represent numbers is introduced. Also, formulas are developed from verbal statements. It is important to impress upon the student the necessity of _learning_ the full meaning of terms as they are introduced. The material in later sections will be confusing and difficult unless the terminology used is well understood.

Examples of algebraic expressions formed by addition, subtraction, multiplication, division, and taking square roots are shown in Section 4-2. Here it is important that the student learn the distinction between _term_ and _factor_; this distinction is very important in later discussions.

Section 4-3 provides a brief introduction to the basic operations in algebra. The operations of combining like terms and the use of the distributive law in multiplication are discussed. The section also describes division of numerator and denominator by common factors (cancellation) for simple algebraic expressions. Again, as in arithmetic, it is important that this operation be properly understood, for in later work in algebraic fractions it is often mishandled by students.

Chapter 5   SIMPLE EQUATIONS AND INEQUALITIES

This chapter is devoted to the solution of simple equations, inequalities, and problems in variation. It also deals with the interpretation of stated problems.

Whereas students easily acquire the skill to solve simple equations of the type in Section 5-1, they find solving the literal equations of Section 5-2 much harder. Concentration on the proper identification of steps to be followed in a given solution is often helpful in overcoming difficulties. For example, when both sides of an equation should be divided by a quantity, some students subtract the quantity. Performing operations with symbols tends to be somewhat more troublesome than with explicit numbers.

In Section 5-3 simple inequalities are considered. The meaning of a solution to an inequality and the operation of multiplying each side by a negative number are points to emphasize. If time limitations require it, this section may be omitted without loss of continuity of coverage.

In Section 5-4 certain basic procedures are outlined for the solution of stated problems. Students often complain about their inability to solve word problems. One of the reasons why solving stated problems is difficult is that it is not possible to follow a fixed procedure for setting up the equations. Therefore, the student must examine the statement of the problem very closely for useful information. Reading and interpreting the problems are the primary sources of confusion. The student should be encouraged to make a systematic analysis of each stated problem according to the outline in this section and not to attempt guesswork. It is quite probable that more than one class session will be necessary for this section.

Ratio and proportion are covered in Section 5-5. An understanding of these concepts is important to the development of variation in the following section, as well as in many applications.

In Section 5-6 emphasis should be placed on setting up equations for the different types of variation. Establishing equations by evaluating k should be emphasized, rather than methods involving proportions. Although the proportion method seems very applicable for direct variation, it confuses the student in the other forms of variation. The instructor should point out that useful scientific equations are often determined by obtaining a set of experimental values and finding k.

In variation problems, care must be taken to distinguish the set of values given to determine k from the set given to find the desired value. Also, the proper product of quantities in combined variation should be established carefully.

8

Chapter 6   INTRODUCTION TO GEOMETRY

A review of basic geometric figures and concepts is given in this chapter. The figures and formulas discussed are fundamental and important in the development of mathematics.

Nearly all of the geometric concepts, figures, and formulas in this chapter should have been previously encountered in high school. However, students usually do not recall enough to allow them to use the figures and formulas competently. Most difficulties arise because the student has not spent sufficient time in assimilating the material. Again, it is essential that all basic facts and concepts be learned thoroughly. Partial learning is of little value.

In Section 6-1 basic geometric figures are defined and identified. The concept of perimeter is discussed in Section 6-2. The student should be encouraged to find perimeters from the definition of perimeter, where possible. This technique is better than developing specific formulas for each figure.

The concept of area is the principal topic of Section 6-3. Again, the main source of difficulty arises from only partially learning the basic formulas.

The concept of volume is the topic of Section 6-4. Volumes of rectangular solids, cubes, and spheres are included.

Additional topics of geometry are developed in Chapter 15.

Chapter 7   BASIC ALGEBRAIC OPERATIONS

This chapter continues the development of the basic algebraic operations begun in Chapter 4.

In Section 7-1, algebraic addition and subtraction are discussed. The operation which should be stressed is the removal of grouping symbols. Here it cannot be overemphasized that the sign of all terms within the grouping are changed if the grouping symbol is preceded by a minus sign. A very common error is to change the sign of the first term only.

In Section 7-2, the multiplication and division of monomials are considered. The distinction between the expressions $a^m \cdot a^n$ and $(a^m)^n$ should be stressed.

Formulas 7-1 through 7-4 must be mastered. Improper handling of exponents and minus signs causes most of the problems in multiplication.

Section 7-3 covers the operations involving multinomials. When multiplying a multinomial by a monomial, we should stress that <u>every</u> term of the multinomial is multiplied by the monomial. Also, when multiplying two multinomials, we should stress that we must multiply <u>each</u> term of one by each term of the other.

In Section 7-4 we consider division involving multinomials. In Example A we should carefully note that each
term of the numerator is divided by the denominator. Most
problems arise in the division of multinomials by monomials. It must be carefully pointed out that each term of
the numerator is to be divided by the monomial. Also, improper handling of signs causes some difficulty.

Chapter 8   FACTORING

Factoring is basic to the material of Chapter 9 (algebraic
fractions) and Chapter 11 (quadratic equations). It should
be impressed upon students who question the need for this
topic that the development of other mathematical concepts
is dependent upon factoring.

Section 8-1 covers the meaning of factoring and the removal of common monomial factors. Persuade the students to
look for such monomial factors before they proceed with the
other factoring steps covered in sections which follow.
Most expressions are more easily factored if this procedure is used.

Factoring the difference of two squares, as discussed
in Section 8-2, depends primarily on the recognition of the
squared quantities. To factor trinomials (Section 8-3) the
student must determine the coefficients of the literal sym-

11

bol and of the constants which properly give rise to the middle term. A complete understanding of the process of multiplication of binomials is of great assistance in this determination.

In Section 8-4, factoring the sum or difference of two perfect cubes depends on the recognition of the cubed quantities. If constraints require it, this section could be omitted without loss of continuity of coverage.

It should be emphasized that, after one step of factoring has been properly accomplished, each resulting factor should be checked for factorability. Until all factors are prime, the process of factoring is not complete.

Only a few of the many basic types of factoring problems are introduced in this chapter. However, they should be sufficient for the elementary problems which are encountered.

Chapter 9   ALGEBRAIC FRACTIONS

The purpose of this chapter is to develop the basic operations on algebraic fractions. Solving equations involving fractions is also discussed. Throughout the development of this chapter, the need for practice should be emphasized.

Section 9-1 shows that, as in arithmetic, equivalent fractions can be formed by multiplying or dividing both numerator and denominator of the fraction by the same quantity. The student should pay very close attention to the special note on cancellation (page 267) and to Example G which follows it. Improper cancellation is a very common and very serious error in beginning algebra. It cannot be overemphasized that cancellation is a process of _dividing_ numerator and denominator by the same factor.

Students often have difficulty in comprehending Eq. (9-1) in Section 9-1. The method of changing a factor to its negative should be covered very carefully and practiced by the student.

Multiplication and division of fractions, as shown in Section 9-2, are generally handled reasonably well by most students. Any difficulties that may arise here are probably caused by improper cancellation.

In addition and subtraction of fractions (Section 9-4), the biggest single difficulty is usually determining the lowest common denominator. This is why an entire section (Section 9-3) is devoted to this topic alone. Once students are able to handle this step, the only difficulties usually encountered are those of cancellation.

Section 9-5 concerns equations involving fractions. The point which must be emphasized here is that of multiplying each term in the equation by the lowest common denominator to clear fractions. Some students have difficulty in actually performing this step.

13

Chapter 10   EXPONENTS, ROOTS, AND RADICALS

This chapter supplements the early material on exponents by covering zero, negative, and fractional exponents. Also, the basic operations on radicals are developed.

In Section 10-1, careful attention must be paid to the meaning of negative exponents as shown in Eq. (10-1). The fact that only the <u>sign</u> of the exponent is changed when a <u>factor</u> is moved to the denominator or to the numerator should be stressed.

Writing numbers in scientific notation is covered in Section 10-2. The convenience and preciseness of notation should be emphasized. Also, its use in the form of metric prefixes is of importance. In Chapter 12, scientific notation is important in the use of logarithms.

One point which should be emphasized in Section 10-3 is that the principal square root of a positive number is positive, not both a positive number and a negative number. Also, it should be stressed that the principal cube root of a negative number is negative and that the square root of a negative number is imaginary. It is a common error to confuse these roots of negative numbers.

Simplifying radicals is the topic of Section 10-4. One problem arises from the fact that students often have difficulty in recognizing one number as the product of a perfect square and another number. Knowing the basic perfect square helps in this determination. Another problem arises in recognition of perfect square factors of higher powers of literal numbers.

Many students encounter some difficulty in rationalizing denominators, especially if a higher power of a literal symbol appears in the denominator. Also, simplification of roots other than square roots can cause problems, again because of failure to recognize perfect powers of numbers and literal symbols. Therefore, Section 10-4 should be covered carefully, and extra class time may be needed.

In Section 10-5, be sure your students understand that when adding and subtracting radicals, all radicals must be put in simplest form before combining. Students generally do not have trouble with multiplication of radicals or with rationalization of radicals of the type in Example H, probably because a set procedure can be followed. However, the instructor should show why this rationalization technique is followed.

Section 10-6 points out that a fractional exponent is equivalent to the root of a number. It should be emphasized, however, that in evaluating numerical expressions involving fractional exponents, the student should first find the root (indicated by the denominator of the fractional exponent) and then raise this result to the power indicated in the numerator of the exponent. That is, we

15

should treat $a^{m/n}$ as $(a^{1/n})^m$ and not as $(a^m)^{1/n}$.
This method enables us to find the root of the smaller number. A knowledge of this section is important to the development of logarithms in Chapter 12.

Chapter 11   QUADRATIC EQUATIONS

This chapter presents basic methods of solving quadratic
equations. Solution by factoring and by use of the quadratic formula is emphasized.

Section 11-1 defines the quadratic equation and ident-
ifies its proper form. Students must learn to determine
accurately the constants a, b, and c in order to use the
quadratic equation properly in Section 11-3. The most difficult constants to recognize are those which are negative,
have a value of 1, or have a value of 0. Emphasize that
the equation should be in proper form before the constants
can be determined.

Solving quadratic equations by factoring is covered in
Section 11-2. It should be pointed out to the student that
the factors themselves are not the solutions but that the
solutions are found by setting the factors equal to zero.

The quadratic formula is discussed in Section 11-3.
The principal purpose here is to have the students learn to
use the formula directly, primarily a matter of identifica-

16

tion of a, b, and c and of proper substitution. The main

problem encountered involves proper handling of signs. The

quadratic formula is derived on pages 354-355. The method

of completing the square is used to derive the quadratic

formula. However, solving quadratic equations by complet-

ing the square is not emphasized. Beginning students in

algebra often have trouble in understanding the method, and

it is not necessary for any other material in this text.

Chapter 12   LOGARITHMS

The concept of a logarithm and its basic uses are intro-

duced in this chapter. The importance of logarithms in

definitions and applications are covered along with the

basic properties.

Section 12-1 includes the basic definition of a loga-

rithm and covers the conversion between the logarithmic and

exponential forms. The basic meaning of a logarithm is

emphasized in this section.

In Section 12-2 we consider the properties of loga-

rithms and how to use these properties for finding pro-

ducts, quotients, powers, and roots of numbers. Care must

be taken with forms of logarithms which combine operations

in a calculation. The importance of doing calculations by

logarithms in order to learn their properties should be

noted even though such calculations can readily be performed on calculators.

Natural logarithms are covered in Section 12-3. It should be emphasized that although the base is different, the properties are the same. Also, the importance of natural logarithms in applications, not calculations, should be noted.

In Section 12-4 the method of interpolation is discussed. In addition to its relevance to logarithms, this method is important for its use with a variety of different circumstances. It should be stressed that even though calculators can be used to avoid interpolation with logarithms, many other mathematical and physical tables require the general interpolation technique presented in this section.

For those instructors who wish to develop logarithms for simple calculations, only Sections 12-1 and 12-2 should be covered in detail. Sections 12-3 or 12-4 may be omitted if the instructor does not care to cover natural logarithms or interpolation. Also, if it is considered unnecessary to cover the topic of logarithms, the chapter may be omitted without loss of continuity.

Chapter 13   GRAPHS

Basic graphical methods are presented in this chapter. The associated topics of functions and rectangular coordinates are also discussed.

Section 13-1 gives a brief introduction to functions. The basic purpose of this section is to acquaint the student with functional notation and with evaluating expressions using signed numbers. A principal problem often encountered is the evaluation when negative numbers are used.

The rectangular coordinate system is introduced in Section 13-2. The student should learn the terminology and concepts of this section, because they are used in the last three sections of this chapter and in Chapter 14.

In Section 13-3, graphs of straight lines and parabolas are considered. The student is shown the basic form of the linear function and can see that its graph is a straight line. Graphing the linear function by intercepts is emphasized. Once the student understands the meaning of an intercept, this graph should be done easily.

In graphing the parabola, in Section 13-3, one of the basic difficulties encountered is evaluating the function for negative values of x. Extra practice may be needed here. Also, the student should be encouraged to set up tables so that the values of x increase. This will help avoid the pitfall of joining points out of order.

Graphs of simple forms of other functions are shown in Section 13-4. The comments regarding the plotting of points for the parabola are also valid for these graphs. An additional problem may occur for evaluations near values of x for which the function is undefined. This section may be omitted if only a basic introduction to graphing is desired.

Section 13-5 covers the important technical application of reading values from a graph. Also, the method of solving equations graphically is shown. This section provides additional practice in constructing and reading graphs.

Section 13-6 introduces graphs involving inequalities. The content of Section 13-3 must be mastered before this section can be covered. The importance of determining the correct region should be emphasized.

In Section 13-7, the circle graph (or pie chart), broken line graph, bar graph, and histogram are discussed.

For instructors with limited time, Sections 13-1 through 13-4 can be used as a basis for introducing the fundamentals of graphing and the remaining sections can be omitted without loss of continuity.

Chapter 14   SIMULTANEOUS LINEAR EQUATIONS

In this chapter we cover methods of solving systems of two linear equations in two unknowns and (in Section 14-5) methods of solving systems of three linear equations with three unknowns. Four methods - graphical, substitution, addition-subtraction, and determinants are shown. The last section contains stated problems which lead to systems of simultaneous linear equations.

Section 14-1 shows the basic graphical method of solving a system of two linear equations. This method is included primarily to give the student a visual conception of the solution. It should be emphasized that the coordinates of the point of intersection give the required solution and that graphing the lines is only a technique which leads to the solution. Students should graph the straight lines by use of intercepts.

Algebraic substitution is important because it is a basic method of solving a system of equations, whether or not they are linear. Algebraic manipulation, especially with fractions, is the principal source of difficulty in Section 14-2. The addition-subtraction method of solving simultaneous equations is shown in Section 14-3. Most students can manipulate this method with great success and will thus attempt to avoid the graphical and substitution methods. Therefore, in testing it may be advisable to specify the method to be used for a solution. One difficulty that might arise in using the addition-subtraction method is that some students neglect to multiply both sides of an equation in the process of solving the system.

In Section 14-4, second-order determinants are defined

and used to solve systems of two equations with two unknowns (Cramer´s rule). It should be noted that this method is especially useful for systems involving messy numbers, and it is the method often used in computer programs.

Section 14-5 covers algebraic methods (not graphical and not determinants) for solving systems of three equations with three unknowns. The methods of substitution and addition-subtraction are extended to this case.

Dependent and inconsistent systems are mentioned in this chapter, but they should not be emphasized because students often have difficulty in understanding their significance.

The exercises of the first five sections of this chapter contain stated problems which result in simultaneous equations. In Section 14-6 the setting up of these equations is stressed. Refer students back to Section 5-4 for suggestions in analyzing stated problems.

Chapter 15   ADDITIONAL TOPICS FROM GEOMETRY

This chapter deals with a number of topics from geometry, such as basic figures, plane and solid, terminology, the Pythagorean theorem, and the concept of similar figures.

Section 15-1 is devoted to various types of angles, and Section 15-2 covers basic properties of triangles, quadrilaterals, and circles. These sections give the student a good working vocabulary in geometry and provide material basic to the following sections.

The Pythagorean theorem is covered in Section 15-3. Be sure to emphasize the overall importance of this topic in all of mathematics and applications.

Section 15-4 concerns similar triangles. Also included is a brief discussion of similar and congruent figures in general. The basic properties of similar triangles are important, for they form the basis for developing trigonometry. The important technical application of scale drawings is also covered in this section.

The basic solid geometric figures are discussed in Sections 15-5 and 15-6. Formulas for volume, lateral area and total area are given for most of the figures. The principal source of difficulty is insufficient mastery of the basic formulas. Formulas for surface area are often confused with those for volume.

Chapter 16   TRIGONOMETRY OF RIGHT TRIANGLES

This chapter introduces the student to the subject of trigonometry. The coverage includes the definitions of the six ratios in terms of the sides of a right triangle, the solution of right triangles, and basic applications.

The six trigonometric ratios are defined in Section

23

16-1. The fact that ratios of corresponding sides of similar triangles are the same is emphasized in this section. Finding values of all the trigonometric ratios of an angle for which one ratio is known causes difficulty for some students.

Section 16-2 covers the use of the table of trigonometric ratios. The student will learn how to find values of the ratios for specified angles and to determine an angle from a given ratio. The use of a calculator is also discussed.

The solution of a right triangle, as well as elementary applications, are discussed in Section 16-3. Students have some difficulty in basic triangle solution, and as with most stated problems, the greatest difficulty arises in setting up the solution.

Chapter 17   TRIGONOMETRY WITH ANY ANGLE

This chapter covers the trigonometric functions of general angles and the solution of oblique triangles. The methods of solution which are used are the Law of Sines and the Law of Cosines.

The signs of the trigonometric functions in each of the four quadrants are discussed in Section 17-1. The importance of knowing these signs for later use should be emphasized.

The determination of trigonometric functions of angles of any magnitude is discussed in Section 17-2. Careful use of the reference angle, and the proper sign of the given function, are important to finding the value of the trigonometric functions of angles in the second, third, and fourth quadrants.

The Law of Sines is covered in Section 17-3. The student must clearly understand which types of combinations of given information may be used with the Law of Sines. Also, the fact that the Law of Sines is actually three equations combined should be noted. The ambiguous case is discussed, but it may be omitted without loss of continuity.

The Law of Cosines is covered in Section 17-4. Students often have difficulty using the Law of Cosines when given the three sides of a triangle. The use of the Law of Sines to complete a solution should be noted.

For courses in which trigonometric graphs are important, but triangle solution is not important, Sections 17-1 and 17-2 should be covered in preparation for the graphing in Chapter 19. The Laws of Sines and Cosines may be omitted without loss of continuity.

Chapter 18   VECTORS

This chapter introduces the concept of a vector. Coverage includes the elementary vector operations of addition, sub-

traction, and scalar multiplication by using both graphic
and analytic techniques.

Section 18-1 discusses vectors, scalars, and displacement. In this section, only graphic techniques are used to
combine vectors. The graphic techniques used are the tail-
to-head method and the parallelogram method.

Section 18-2 uses right triangle trigonometry to resolve vectors into rectangular components. It should be
stressed that the angle used in the calculation is the
angle in standard position.

Section 18-3 involves an analytic method of adding vectors. The vectors are first resolved into rectangular components. The x components are combined into one vector and
the y components are combined into another vector. The
magnitude of the resultant is found by using the Pythagorean theorem with those x and y components. It should be
stressed that the resultant vector has not been determined
until both its magnitude and direction are known. A variety of applications are included. Students usually learn
to resolve a vector into its components reasonably well,
but they often have difficulty in adding vectors by first
resolving them into components. This topic may well require more than one lesson.

Chapter 19   RADIANS AND TRIGONOMETRIC CURVES

In this chapter radians and the graphs of the sine and cosine are covered. Radians are introduced for use in graphing, and direct applications of radians are also noted.

Radians are introduced in Section 19-1. The use of
radians in terms of pi, and in terms of general numbers is
covered. Students must clearly understand the relationship
between radians and angles in degree measure in order to
avoid difficulty. Applications of radians are covered in
Section 19-2. This section may be omitted without loss of
continuity if graphing is the primary goal in this chapter.

Graphs of the sine and cosine functions are introduced
in Section 19-3. The basic type of graph and amplitude are
the principal topics of this section. In Section 19-4
these graphs are sketched with the concept of period being
introduced. In Section 19-5 these graphs are sketched with
the concept of displacement being used.

The student should be encouraged to sketch, rather than
plot, these graphs. The use of displacement along with
period often causes difficulty. This chapter may be stop-
ped after any given section, depending on the intent of the
course.

Chapter 20   COMPLEX NUMBERS

In this chapter we introduce complex numbers and consider
their arithmetic operations of addition, subtraction, multiplication, and division. We also represent them graphically in the complex plane, and convert between rectangular
and polar forms.

In Section 20-1 we introduce the rectangular forms of
complex numbers. We illustrate methods for converting
complex numbers expressed with radicals to the standard
rectangular form in terms of j. Students tend to make
mistakes by incorrectly applying Eq. (10-8) so it would be
helpful to stress Example C on page 640 and the comments
which follow it.

In Section 20-2 we present the basic operations of
arithmetic. Students usually have little difficulty with
addition and subtraction. With multiplication we should
emphasize that we are really multiplying two binomials (as
discussed in Section 7-3). It is therefore essential that
each term of the first factor must be multiplied by each
term of the second factor. This reinforces the multiplication process discussed in Section 7-3. As expected, division is the operation that causes most difficulty. It is
helpful to point out that the division of complex numbers
involves the same approach used in rationalizing factors in
which the denominator consists of two terms, at least one

of which is a radical. Point out the similarity between the use of conjugates in Example H on page 327 and Example F on page 645.

Section 20-3 includes the complex plane and the graphic representation of complex numbers. This representation allows us to treat complex numbers as vectors. Example E on page 650 provides a good application of this concept.

In Section 20-4 we show how to convert complex numbers between the rectangular and polar forms. Many applications require complex numbers expressed in polar form. Because of the nature of the polar form, this section tends to reinforce some of the work done earlier with vectors in Chapter 18. The student is reminded that the polar form requires both magnitude and direction. Also, we again consider the expression of the reference angle so that it corresponds to the correct quadrant.

If you need a quick introduction to the basics of complex numbers, cover only Sections 20-1 and 20-2.

Appendix A: THE SCIENTIFIC CALCULATOR

This appendix is primarily intended to be a reference, but instructors may wish to use it at the beginning of the course. A brief description of the nature and importance of the algebraic operating system will be especially help-ful to students who intend to acquire a new calculator.

Appendix B: BASIC PROGRAMMING

If the facilities are available, computer usage makes an excellent supplement to this course. Students with ade-quate programming backgrounds might be encouraged to de-velop some original programs while others might be en-couraged to enter and use some of the programs listed.

Appendix C: NUMBERS IN BASE TWO

This topic is a brief introduction to numbers in another base. Through addition and multiplication in base two (often associated with computers), the student will gain further insight into the basic arithmetic operations.

COMPARISON WITH BASIC TECHNICAL MATHEMATICS, 5th ed.

In the table on the following two pages, the material covered in Introduction to Technical Mathematics, 4th ed., is compared to Basic Technical Mathematics, 5th ed. The table shows the major areas of coverage and may aid the instructor in choosing texts appropriate for courses under consideration.

In this table, Introduction to Technical Mathematics, 4th ed. (ITM), and Basic Technical Mathematics, 5th ed. (BTM), are designated as noted. In designating material covered, X indicates a basic coverage, R indicates that the coverage is review in nature, and A indicates that the topic is covered in the appendix.

Topic	ITM	BTM
Fundamentals from arithmetic	X	
Measurement and approximate numbers	X	A
The calculator	X	X
The metric system	X	A
Signed numbers	X	R
Basic geometric concepts and figures	X	A
Additional topics from geometry	X	A
Introduction to algebra	X	R
Simple equations	X	R
Basic algebraic operations	X	X
Factoring	X	X
Algebraic fractions	X	X
Exponents, roots, and radicals	X	X
Quadratic equations	X	X
Logarithms	X	X
Exponential functions		X
Functions and graphs	X	X
Simultaneous linear equations	X	X
Determinants	X	X
Right angle trigonometry	X	X
Radians	X	X
Trigonometric functions of any angle	X	X

32

Topic	ITM	BTM
Vectors	X	X
Oblique triangles	X	X
Graphs of the trigonometric functions	X	X
Complex numbers (the j-operator)	X	X
Systems of non-linear equations		X
Higher-degree equations		X
Matrices		X
Introduction to inequalities	X	X
Additional topics with inequalities		X
Variation	X	X
Progressions		X
Binomial theorem		X
Fundamental trigonometric relations		X
Trigonometric equations		X
Inverse trigonometric functions		X
Analytic geometry		X
Polar coordinates		X
Statistics		X
Curve fitting		X

Note: In many of the topics listed, the coverage is more extensive in BTM than in ITM.

ANSWERS TO EXERCISES

EXERCISES 1-1

1. 152	2. 171	3. 1234	4. 1154
5. 2420	6. 2461	7. 21,410	8. 20,859
9. 25,295	10. 19,045	11. 139,639	12. 340,851
13. 5602	14. 3121	15. 581	16. 697
17. 949	18. 809	19. 9798	20. 9989
21. 30,579	22. 978	23. 95,682	24. 189,786
25. 8 ft 2 in.	26. 12 ft 6 in.	27. 18 lb	28. 3 lb 4 oz
29. 5 ft 5 in.	30. 2 ft 11 in.	31. 4 lb 11 oz	32. 1 lb 7 oz
33. 10 ft 3 in.	34. 13 ft 4 in.	35. 11 ft 10 in.	36. 24 ft 11 in.
37. 29 yd 1 ft	38. 31 ft	39. 25 ft 4 in.	40. 26 ft 7 in.
41. 11 in.	42. 1 ft 10 in.	43. 4 ft 8 in.	44. 1 ft 10 in.
45. 8 yd 2 ft	46. 8 ft 5 in.	47. 10 in.	48. 7 ft 6 in.
49. 530 ft	50. 1140 gal	51. $1691	52. 9 min 37 s
53. 4 h 41 min 30 s			54. 11 lb 2 oz
55. 278 gal	56. 66 Ω	57. 156 ft	58. 218 g
59. 10 h 43 min			60. 5 gal 2 qt
61. 188 kg	62. $37	63. 16,250 mi^2	64. 3668 lb

EXERCISES 1-2

1. 10,534	2. 22,545	3. 1,314,976	4. 262,144
5. 1,048,576	6. 16,008,876		
7. 236,894,403	8. 448,962,514		
9. 244	10. 39	11. 324	12. 508
13. 2048	14. 2228, rem 3		
15. 1981, rem 104	16. 1267, rem 310		
17. 47,804	18. 209,664	19. 3,183,390	20. 247,104
21. 150	22. 280	23. 72,220	24. 5,069,821
25. 160 mi^2	26. 1056 ft^2	27. 238 in.^{2}	28. 186,186 ft^2
29. 10,360 cm^2	30. 446,346 ft^2		
31. 432 in.^{2}	32. 1224 in.^{2}	33. 10,115 mi	34. 4680 r
35. $1280	36. 108 ft	37. 330 ft	38. 572 m
39. 18 mi/gal	40. $35	41. 40 mi	42. 1125 mi/h
43. $5690	44. 67,664 m^2	45. 48	46. 552,000
47. No	48. 1186		

EXERCISES 1-3

1. $\frac{5}{9}$ 2. $\frac{4}{11}$ 3. $\frac{1}{7}$ 4. $\frac{8}{8}$

5. $\frac{7}{13}$; $\frac{3}{16}$ 6. $\frac{1}{3}$; $\frac{11}{17}$ 7. $\frac{9}{8}$; $\frac{1}{12}$ 8. $\frac{23}{4}$; $\frac{2}{35}$

9. 1; 6   10. 1; 19   11. 32; 1   12. 503; 1

13. $1\frac{2}{3}$ 14. $3\frac{3}{5}$ 15. $4\frac{12}{13}$ 16. $1\frac{23}{32}$

17. $3\frac{53}{75}$ 18. $9\frac{27}{32}$ 19. $53\frac{4}{25}$ 20. $37\frac{10}{118}$

21. $\frac{17}{5}$ 22. $\frac{27}{4}$ 23. $\frac{79}{8}$ 24. $\frac{63}{5}$

25. $\frac{223}{13}$ 26. $\frac{275}{8}$ 27. $\frac{423}{4}$ 28. $\frac{5891}{25}$

29. $\frac{17}{24}$ 30. $\frac{7}{12}$ 31. $\frac{10}{23}$ 32. $\frac{55}{48}$

33. $8\frac{1}{2}$ in.   34. $9\frac{3}{4}$ oz   35. $3\frac{3}{4}$ gal   36. $\frac{21}{4}$ in.

37. $\frac{37}{10}$ in.   38. $\frac{517}{100}$ in.   39. $\frac{3}{8}$ 40. $\frac{7}{10}$

EXERCISES 1-4

1. $\frac{6}{14}$ 2. $\frac{15}{27}$ 3. $\frac{4}{5}$ 4. $\frac{3}{25}$

5. $\frac{24}{78}$ 6. $\frac{88}{165}$ 7. $\frac{5}{13}$ 8. $\frac{20}{6}$

9. $\frac{91}{175}$ 10. $\frac{204}{180}$ 11. $\frac{32}{2}$ 12. $\frac{17}{20}$

13. $\frac{1}{2}$ 14. $\frac{2}{3}$ 15. $\frac{3}{2}$ 16. $\frac{7}{2}$

17. $\frac{4}{5}$ 18. $\frac{3}{4}$ 19. $\frac{3}{5}$ 20. $\frac{2}{5}$

21. $\frac{2}{5}$　　22. $\frac{5}{6}$　　23. $\frac{2}{5}$　　24. $\frac{3}{26}$

25. 4　　26. 10　　27. 15　　28. 24
29. 3　　30. 12　　31. 4　　32. 8
33. 40　　34. 3　　35. 16　　36. 64
37. 2×2×5　　　　　　　　38. 2×2×7
39. 2×2×2×2　　　　　　　40. 2×2×2×2×2
41. 2×2×3×3　　　　　　　42. 2×2×11
43. 2×2×2×2×3　　　　　　44. 2×2×13
45. 3×19　　　　　　　　　46. 2×2×3×7
47. 3×5×7　　　　　　　　　48. 2×2×3×11

49. $\frac{6}{7}$　　50. $\frac{4}{9}$　　51. $\frac{4}{5}$　　52. $\frac{3}{5}$

53. $\frac{7}{3}$　　54. $\frac{11}{3}$　　55. $\frac{2}{3}$　　56. $\frac{5}{6}$

57. $\frac{1}{4}$　　58. $\frac{1}{3}$　　59. $\frac{3}{5}$　　60. $\frac{2}{3}$

61. $\frac{3}{4}$ in.　　62. $\frac{24}{64}$ in.　　63. $\frac{7}{16}$　　64. $\frac{53}{60}$

65. $\frac{3}{10}$　　66. $\frac{8}{15}$　　67. $\frac{2}{5}$　　68. $\frac{13}{19}$

69. $\frac{1}{5}$　　70. $\frac{19}{100}$　　71. $\frac{16}{31}$　　72. No ($\frac{1}{64}=\frac{4}{256}$)

EXERCISES 1-5

1. 4　　2. 6　　3. 24　　4. 18
5. 72　　6. 210　　7. 300　　8. 264
9. $\frac{4}{5}$　　10. $\frac{7}{11}$　　11. $\frac{2}{7}$　　12. $\frac{3}{5}$

13. $\frac{3}{4}$　　14. $\frac{1}{3}$　　15. $\frac{1}{8}$　　16. $\frac{1}{2}$

17. $\frac{13}{12}$　　18. $\frac{43}{30}$　　19. $\frac{1}{10}$　　20. $\frac{11}{63}$

21. $\frac{7}{3}$　　22. $\frac{1}{4}$　　23. $\frac{47}{40}$　　24. $\frac{29}{63}$

25. $\frac{28}{39}$ 26. $\frac{79}{168}$ 27. $\frac{83}{24}$ 28. $\frac{103}{36}$

29. $\frac{29}{42}$ 30. $\frac{61}{66}$ 31. $\frac{73}{72}$ 32. $\frac{794}{225}$

33. $\frac{73}{42}$ 34. $\frac{91}{135}$ 35. $\frac{13}{5}$ 36. $\frac{1002}{175}$

37. $\frac{1}{2}$ in.   38. 2 ft   39. $9\frac{1}{2}$ Ω   40. $7\frac{19}{20}$ gal

41. $26\frac{2}{3}$ ft   42. $5\frac{7}{12}$ ft   43. $1\frac{17}{24}$ in.   44. $\frac{2}{15}$

45. $\frac{5}{24}$ 46. $\frac{5}{9}$ 47. $1\frac{23}{32}$ in.   48. $3\frac{1}{16}$ in.

EXERCISES 1-6

1. $\frac{6}{77}$ 2. $\frac{21}{40}$ 3. $\frac{6}{5}$ 4. $\frac{10}{21}$

5. $\frac{21}{20}$ 6. $\frac{35}{8}$ 7. $\frac{5}{27}$ 8. $\frac{15}{16}$

9. $\frac{5}{18}$ 10. $\frac{5}{28}$ 11. $\frac{3}{2}$ 12. $\frac{2}{5}$

13. $\frac{14}{9}$ 14. $\frac{9}{16}$ 15. $\frac{2}{17}$ 16. $\frac{9}{5}$

17. 2   18. $\frac{3}{4}$ 19. $\frac{2}{15}$ 20. $\frac{4}{5}$

21. $\frac{11}{4}$ 22. $\frac{12}{25}$ 23. $\frac{27}{64}$ 24. $\frac{2}{7}$

25. $\frac{168}{23}$ 26. $\frac{1}{12}$ 27. $\frac{5}{4}$ 28. $\frac{12}{7}$

29. 7   30. $\frac{5}{52}$ 31. $\frac{188}{351}$ 32. $\frac{16}{9}$

33. $\frac{57}{4}$ 34. $\frac{25}{132}$ 35. $\frac{60}{209}$ 36. $\frac{19}{87}$

37. $\frac{1}{5}$; $\frac{1}{13}$ 38. $\frac{1}{2}$; $\frac{1}{9}$ 39. 2; 5   40. $\frac{9}{2}$; $\frac{7}{3}$

41. $\frac{3}{16}$; $\frac{2}{7}$ 42. $\frac{4}{13}$; $\frac{32}{289}$ 43. $\frac{16}{81}$; $\frac{10}{71}$ 44. $\frac{10}{121}$; $\frac{3}{46}$

45. $8\frac{1}{2}$ in.   46. $\frac{1}{4}$ in.²   47. 12   48. 45 V

49. $\frac{9}{26}$ acre/h 50. 16   51. $8\frac{4}{5}$ qt   52. $12\frac{22}{25}$ ft/s²

53. $\frac{2}{5}$ in.   54. 24 kg   55. $87\frac{3}{11}$ mi/h 56. $8\frac{3}{4}$ turns

EXERCISES 1-7

1. $4(10)+7(1)+\frac{3}{10}$ 2. $2(10)+9(1)+\frac{2}{10}+\frac{6}{100}$

3. $4(100)+2(10)+9(1)+\frac{4}{10}+\frac{8}{100}+\frac{6}{1000}$

4. $5(1000)+2(100)+3(10)+0(1)+\frac{3}{10}+\frac{7}{100}+\frac{2}{1000}+\frac{7}{10,000}$

5. 27.3   6. 404.88   7. 57.54   8. 6.383
9. 8.03   10. 0.41   11. 17.4   12. 86.302
13. 0.4   14. 0.4375   15. 0.21   16. 0.48
17. 1.7   18. 13.2   19. 0.499   20. 0.068

21. $\frac{4}{5}$ 22. $\frac{1}{500}$ 23. $\frac{9}{20}$ 24. $\frac{3}{40}$

25. $\frac{267}{50}$ 26. $\frac{88}{5}$ 27. $\frac{63}{2500}$ 28. $\frac{21}{2500}$

29. 31.295   30. 210.595   31. 2817.256   32. 206.0902
33. 8.763   34. 0.669   35. 0.02628   36. 0.7698
37. 13.99952   38. 306.9418   39. 9.882   40. 0.080983
41. 124.992   42. 4.002684   43. 0.1215   44. 0.6675

45. 12.5   46. 0.9   47. 295.4   48. 56.2
49. 5.20   50. 117.07   51. 0.46   52. 0.80
53. 0.6 V   54. 0.17 ft   55. $177.49   56. 0.066 in.
57. 28.5 Ω   58. 42.80 cm   59. 235.88 L   60. 0.4555 cm

61. $8\frac{1}{4}$ ft; 8 ft 3 in.   62. 26,070 mi

63. $298.20   64. 1.8   65. $213.18   66. 20.46 m
67. 2.44 m   68. 8.5

EXERCISES 1-8

1. 0.08   2. 0.78   3. 2.36   4. 4.82
5. 0.003   6. 0.00082   7. 0.056   8. 0.103
9. 27%   10. 9%   11. 321%   12. 2160%
13. 0.64%   14. 0.07%   15. 700%   16. 830%

17. $\frac{3}{10}$ 18. $\frac{12}{25}$ 19. $\frac{1}{40}$ 20. $\frac{1}{125}$

21. $\frac{6}{5}$ 22. $\frac{9}{25,000}$ 23. $\frac{57}{10,000}$ 24. $\frac{7}{5000}$

25. 60%   26. 35%   27. 57.1%   28. 145.5%
29. 22.9%   30. 62.1%   31. 266.7%   32. 128.6%
33. 13   34. 5.98   35. 5.304   36. 378
37. 25%   38. 2%   39. 7.5%   40. 4%
41. 50   42. 14.4   43. 400   44. 1.6
45. $40.80   46. $22.68   47. 32%   48. 50,304 J
49. $17.64   50. 2730 gal   51. $3395.75
          52. 738,584 cm³   53. 35%   54. 9%
55. 0.3%   56. 93%   57. 97%   58. 85%
59. $512   60. $580   61. 66%   62. 11 A
63. 4.48%   64. 2217 mi

EXERCISES 1-9

1. 8^3 2. 4^5 3. 2^4 4. 5^3
5. 3^5 6. 12^3 7. 10^5 8. 6^8
9. 8×8   10. 2×2×2×2×2

11. $3\times3\times3\times3\times3\times3$

12. $4\times4\times4$

13. $7\times7\times7\times7\times7\times7\times7\times7$

14. $10\times10\times10\times10$

15. $5\times5\times5\times5\times5\times5$

16. $6\times6\times6\times6\times6$

17. $(8\times10^3) + (5\times10^2) + (4\times10) + 3$

18. $(3\times10^3) + (5\times10^2) + (2\times10) + 7$

19. $5 + (7\times\frac{1}{10}) + (3\times\frac{1}{10^2}) + (9\times\frac{1}{10^3})$

20. $(5\times10) + 6 + (8\times\frac{1}{10}) + (7\times\frac{1}{10^2}) + (2\times\frac{1}{10^3})$

21. 243	22. 128	23. 64	24. 1296
25. 0.09	26. 0.0144	27. 42.875	28. 0.000512
29. 4	30. 9	31. 11	32. 20
33. 4	34. 2	35. 2	36. 3
37. 0.4	38. 0.9	39. 0.3	40. 1.1
41. 108	42. 1600	43. 4500	44. 3,430,000
45. 891	46. 768	47. 0.637	48. 2.45
49. 70,000 J	50. 2,000,000,000	51. 25	52. 262,144
53. 30 ft	54. $\frac{15}{4}$ s	55. 30 Ω	56. 16 m

57. 600,000,000

58. 22 in.

59. 676,000

60. $10,737,418.24

EXERCISES 1-10

1. 4.41	2. 29.16	3. 15.21	4. 65.61
5. 1.449	6. 2.324	7. 1.975	8. 2.846
9. 5.657	10. 8.660	11. 7.416	12. 9.539
13. 13.78	14. 27.57	15. 65.57	16. 309.8
17. 0.2098	18. 0.8426	19. 0.08062	20. 0.01440
21. 4.24	22. 2.28 in.	23. 2.1 s	24. 240.8 ft
25. 2.53 s	26. 31.1 Hz	27. 75.9 kPa	28. 199 ft/s
29. 5.8 in.	30. 2.1	31. 12.2 ft	32. 28.3 m

EXERCISES 1-11

1. 3.07	2. 2.4	3. 6.44	4. 30.75
5. 56	6. 9	7. 6	8. 54
9. 3	10. 4	11. 8	12. 7
13. 8	14. 81	15. 32	16. 125
17. 1.4166667	18. 0.09722222	19. 0.51461988	20. 0.25
21. 11.916667	22. 10.306818	23. 6.6071429	24. 17.75
25. 4.4	26. 1.8775510	27. 0.9	28. 0.625
29. 120	30. 102	31. 132	32. 40
33. 4.3743571	34. 0.03162278		
35. 0.01	36. 0.93808315		
37. 0.33333333	38. 0.08333333		
39. 0.02222222	40. 0.0625	41. 0.34	42. 2.563
43. 3.456	44. 0.08	45. 39.76 in.^{3}	46. 6.24
47. 0.002 Ω	48. 19.7 ft		
49. 403.2	50. 2,560,000	51. 124.8 g	52. 12.0%

EXERCISES 1-12

1. 13,330	2. 111,032	3. 2779	4. 9867
5. 121.112	6. 1089.086	7. 1876.96	8. 1.9884
9. 4,365,872	10. 638,520,864	11. 428	12. 3962
13. 72.312	14. 0.702642	15. 0.014	16. 314
17. 37.11312	18. 2427.972	19. $\frac{5}{13}$	20. $\frac{5}{8}$
21. $\frac{19}{10}$	22. $\frac{17}{84}$	23. $\frac{71}{60}$	24. $\frac{71}{126}$
25. $\frac{401}{525}$	26. $\frac{181}{270}$	27. $\frac{22}{5}$	28. $\frac{5}{36}$
29. $\frac{11}{6}$	30. $\frac{25}{2}$	31. $\frac{4}{3}$	32. $\frac{14}{17}$
33. $\frac{215}{33}$	34. $\frac{18}{243}$	35. 7	36. 12
37. 0.04	38. 0.12	39. 10	40. 2
41. 48	42. 200	43. 49	44. 876

45. $\frac{24}{7}$; $3\frac{2}{5}$ 46. $\frac{59}{8}$; $1\frac{2}{19}$ 47. $\frac{257}{8}$; $17\frac{4}{7}$ 48. $\frac{227}{4}$; $10\frac{2}{17}$

49. $\frac{3}{5}$ 50. $\frac{5}{9}$ 51. $\frac{2}{5}$ 52. $\frac{4}{5}$

53. $\frac{9}{2}$; $\frac{7}{22}$ 54. $\frac{3}{4}$; $\frac{5}{42}$ 55. $\frac{9}{10}$; $\frac{43}{34}$ 56. $\frac{9}{40}$; $\frac{17}{7}$

57. 0.28125   58. 0.3025   59. $\frac{14}{25}$ 60. $\frac{631}{200}$

61. 0.82; $\frac{41}{50}$ 62. 2.50; $\frac{5}{2}$ 63. 0.0055; $\frac{11}{2000}$

64. 0.000225; $\frac{9}{40,000}$

65. 93.4%   66. 8728%   67. 8%   68. 375%
69. 2.098   70. 2.950   71. 14.44   72. 88.36
73. 6.557   74. 8.775   75. 126.5   76. 0.9381
77. 0.734   78. 30.8   79. 3.06   80. 93.4
81. 4.028   82. 17596.3   83. 5.59   84. 23.3
85. 17.98 in.   86. 55.0 m   87. 29.54 in.   88. $11.08
89. 21.5 ft   90. 26,815 mi   91. 20.6 Ω   92. 333.33

93. $\frac{7}{32}$ in.   94. $6\frac{3}{4}$ ft   95. $\frac{3}{10}$ qt   96. 47 ft 2 in.

97. 84.06 m   98. 15.61 ft   99. 0.008 lb   100. $38.85

101. $58.65   102. 30 months   103. 2.5%   104. 11.9 V

105. 9.5 Ω   106. 72°F   107. $20\frac{9}{16}$ in.   108. 1727°C

EXERCISES 2-1

1. mA; 1 mA = 0.001 A
2. μm; 1 μm = 0.000001 m
3. kV; 1 kV = 1000 V
4. MW; 1 MW = 1,000,000 W
5. kW; 1 kW = 1000 W
6. ns; 1 ns = 0.000000001 s
7. ML; 1 ML = 1,000,000 L
8. cg; 1 cg = 0.01 g
9. megavolt; 1 MV = 1,000,000 V
10. kilovolt; 1 kV = 1000 V
11. microsecond; 1 μs = 0.000001 s
12. kilowatt; 1 kW = 1000 W

13. centivolt; 1 cV = 0.01 V   14. milliohm; 1 mΩ = 0.001 Ω
15. nanoampere; 1 nA = 0.000000001 A
16. picosecond; 1 ps = 0.000000000001 s

17. (a) m^2; (b) ft^2 18. (a) m^3; (b) ft^3

19. (a) m/s; (b) ft/s   20. (a) m/s^2; (b) ft/s^2
21. 4 s   22. 3 h   23. 8 m   24. 2 km
25. 4 cs   26. 7 cL   27. 3 μs   28. 6 cm
29. 8 cg   30. 5 mm   31. 4 kg   32. 9 kW

33. mm^2 34. $in.^3$ 35. mi/gal   36. km/s

37. 1/s   38. s/Ω 39. lb/ft^2 40. ft·lb/s

41. kg·m/s^2 42. N·m   43. A·s   44. J/A·s

EXERCISES 2-2

1. 10   2. 100,000   3. 63,360   4. 32,000

5. 288 $in.^2$ 6. 50,000 mm^3 7. 30.48 cm   8. 17.6 lb
9. 220 qt   10. 60,000,000 mL

11. 23.7 L   12. 10.6 in.   13. 52,000 cm^2 14. 205 cm^3

15. 9.05 ft^3 16. 15.8 L   17. 3.09 L   18. 44.4 km
19. 69.13 kg   20. 521.89 in.   21. 8.29 lb   22. 2.315 gal
23. 1829 mm   24. 83.7 L   25. 90 t   26. 4.25 mi
27. 50 mi/h   28. 3890 lb   29. 3.8 L   30. 21 mm

31. 32.2 ft/s^2 32. 1129 ft/s   33. 1000 kg/m^3 34. 5520 kg/m^3
35. 3270 ft/s   36. 4.54 L/s   37. 0.47 L/s   38. 339 N·m

39. 60.9 kcal   40. 9.753 m/s^2

EXERCISES 2-3

1. exact   2. approx.   3. approx.   4. exact
5. approx.   6. 25 is exact but 79.3 is approximate.
7. Both 1's are approximate but 2.80 is exact.
8. 40 is exact and 987 is approximate.
9. 3; 4   10. 3; 2   11. 4; 4   12. 2; 4

13. 3; 3   14. 1; 5   15. 4; 5   16. 4; 1
17. (a) 3.764; (b) 3.764   18. (a) same; (b) 7.673
19. (a) 0.01; (b) 30.8   20. (a) 70,370; (b) 70,370
21. (a) same; (b) 78.0   22. (a) 37.1; (b) same
23. (a) 0.004; (b) same   24. (a) same; (b) 50.060
25. 5.71; 5.7   26. 53.7; 54
27. 6.93; 6.9   28. 27.8; 28
29. 4100; 4100   30. 287; 290
31. 46800; 47000   32. 32800; 33000
33. 501; 500   34. 7440; 7400
35. 0.215; 0.22   36. 0.635; 0.64
37. 128.25 ft; 128.35 ft   38. 0.7675 in.; 0.7685 in.
39. 81.5 L; 82.5 L   40. 45.5 lbs; 46.5 lbs
41. 0.1733 qt   42. 1.59 mm
43. 91.4 m   44. 24.00 ft
45. Answer varies with calculators.
46. Answer varies with calculators.
47. Most calculators will yield 24 so that the accuracy is reduced to 2.
48. Answer varies with calculators.

EXERCISES 2-4

1. 39.0	2. 305	3. 1.78	4. 363.9
5. 754.0	6. 5.9	7. 2.59	8. 4.34
9. 113	10. 4.02	11. 430	12. 20.41
13. 110	14. 0.000750	15. 790	16. 229.8
17. 6.9	18. 2.9	19. 9.35	20. 0.220
21. 10.2	22. 8.1	23. 22	24. 8.2
25. 17.62	26. 14.882	27. 18.85	28. 0.00499
29. 13.2	30. 7.320	31. 8.1	32. 17.9
33. 23.33	34. 84.7	35. 0.632	36. 3540
37. 62.23	38. 0.095	39. 0.367	40. 220
41. 196 ft	42. 27.0 lb		
43. 262,144 bytes			44. 0.0373 W
45. 64.3 ft/s	46. $577.47	47. 154 m	48. 5.250 in.
49. 5440 ft	50. 784.4 in.	51. 56.76 L	52. 42 km

53. 433 V   54. Tin; 70 kg/m^3
55. Too many significant digits; time has only two significant digits.
56. 7800

EXERCISES 2-5

1. 4;4	2. 1;2	3. 3;4	4. 4;4

5. (a) 7.32 (b) same   6. (a) 80.0 (b) 80.0
7. (a) Same (b) 207.31   8. (a) 0.0021 (b) 98.568
9. 98.5; 98   10. 2.73; 2.7
11. 60500; 61000   12. 220000; 220000
13. 673; 670   14. 69000; 69000
15. 0.700; 0.70   16. 4940; 4900

17. 438.7	18. 29.8	19. 10.89	20. 801
21. 0.12	22. 40,000	23. 0.057	24. 208
25. 6.9	26. 1.74	27. 7.6	28. 2.78
29. 31.44	30. 16.7	31. 29	32. 0.294

33. microgram; 1 μg = 0.000001 g
34. centiampere; 1 cA = 0.01 A
35. ks; 1 ks = 1000 s   36. MV; 1 MV = 1,000,000 V
37. 0.385 cm^3 38. 4750 cm^2 39. 13 cm   40. 1.18 m
41. 35.62 pt   42. 1,850,000 cm
43. 760 L   44. 1.993 mi   45. 740 in.^{3} 46. 0.197 ft^2

47. 1.10 m/s   48. 95,000 cm/h
49. 1.061 t/ft^2 50. 123 kg/m^3

51. 30,630,000 cm^3/min   52. 0.4837 m^3/s
53. 0.1855 in.; 0.1865 in.   54. 1.45 V; 1.55 V
55. $537.50   56. 480 mi/h   57. 21.84 g   58. 0.366 m
59. 0.021 m   60. 6.0 in.
61. 320.2 calc/s   62. 100 V   63. 0.67 kg/L   64. 52 mi/h
65. 0.000287 in.^{2}
66. 0.5 s
67. Cannot obtain difference to tenths, since initial time was not
     recorded to tenths.
68. m/s^2

EXERCISES 3-1

1. Add 3 to 6.	2. Subtract 3 from 6.
3. Add 7 to +2.	4. Subtract +5 from +9.
5. Add -2 to -8.	6. Subtract -1 from +4.

7. Subtract +6 from -3.   8. Subtract -7 from -1.

9, 11, 13, 15:

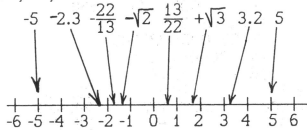

10, 12, 14, 16:

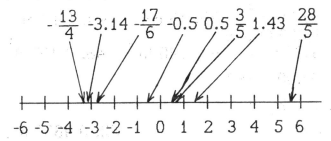

17. 2   18. 0   19. -1   20. -7
21. >   22. >   23. <   24. <
25. >   26. <   27. <   28. >
29. >   30. >   31. =   32. >
33. 6, 6   34. 5, 5   35. $\frac{6}{7}, \frac{8}{5}$ 36. 2.4, 0.1
37. -30   38. -2600   39. - 0.5 V   40. -8
41. The first one. 42. Ans. varies. 43. -37   44. +4.5 cm
45. -30°C < -5°C   46. -60 m > -80 m
47. -2 V > -5 V   48. -0.94 < -0.58
49. Answer varies with calculators.   50. 5 is displayed.
51. Enter 0 - 3 =
52. -85 is a common result.

EXERCISES 3-2

1. +11	2. +11	3. -15	4. -11
5. +3	6. +7	7. -8	8. -8
9. +7	10. +8	11. -3	12. -13
13. -10	14. -11	15. +16	16. +16
17. +1	18. -4	19. -6	20. +8
21. -8	22. -21	23. +1	24. -12
25. +6	26. -8	27. -17	28. -12
29. -2	30. -8	31. -6	32. +9

33. (a) $-5^{\circ}C$ (b) $-15^{\circ}C$ 　　　　　　　　　　　34. -$6

35. 63 cm	36. $150^{\circ}C$	37. 17 A	38. 5850 ft
39. +5000	40. A double negative statement is positive.		
41. -7	42. -5	43. 9	44. 11
45. -3	46. 12	47. 7	48. 44

EXERCISES 3-3

1. +80	2. +60	3. -63	4. -44
5. -84	6. -60	7. +30	8. +72
9. -56	10. -90	11. +240	12. +480
13. +360	14. -420	15. 0	16. +30
17. -168	18. 0		
19. +36; +36	20. +125; -125		
21. +49; -49	22. -81; +81	23. -1024	24. +4096
25. +1; -1	26. -1; - 1		
27. -128; -128	28. -256: +256	29. -50 mm	30. -$12
31. 7000 ft	32. 240 mV	33. 994 lb	34. 224 ft
35. 10 ft	36. Signs alternate: 1, -1, 1, -1, ...		
37. 18	38. -8	39. -56	40. 20
41. -8	42. -8	43. -81	44. 81

EXERCISES 3-4

1. +4	2. +8	3. -9	4. -3
5. -11	6. -5	7. +15	8. +4
9. 0	10. 0	11. Undef.	12. Undef.
13. -16	14. +2	15. +3	16. -4

17. +3   18. -1   19. -5   20. +12
21. -$130   22. 25 days   23. -250 ft   24. 5 s
25. 5 days   26. 3; erect   27. 5250 ft   28. $14,400 net
29. -557   30. -37   gain
31. 8   32. 21   33. -21   34. 2
35. 0   36. Error

EXERCISES 3-5

1. 20	2. 21	3. -14	4. -3
5. -49	6. -35	7. -1	8. -90
9. -69	10. -6	11. -5	12. -2
13. -106	14. -28	15. -20	16. -116
17. 14	18. -60	19. -101	20. 29
21. 19	22. 19	23. 3	24. -11
25. -12	26. -17	27. -2	28. -53

29. -1000 ft   30. -20°F   31. -78.0 m   32. 80.6
33. 46.2 gal   34. 1600 ft^2 35. 2500 gal   36. 1.8 mA
37. 16   38. 18   39. -9   40. 3
41. 22.5   42. -311.94444
43. -0.75   44. -663

EXERCISES 3-6

1. $<$ 2. $>$ 3. $>$ 4. $<$

5. $-\frac{5}{4}$ 2.5   6. $-\frac{2}{9}$ $\frac{3}{7}$

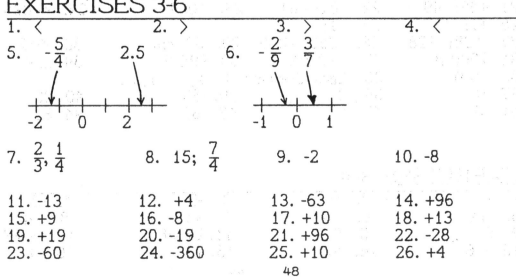

7. $\frac{2}{3}, \frac{1}{4}$ 8. 15; $\frac{7}{4}$ 9. -2   10. -8

11. -13   12. +4   13. -63   14. +96
15. +9   16. -8   17. +10   18. +13
19. +19   20. -19   21. +96   22. -28
23. -60   24. -360   25. +10   26. +4

27. -6	28. -6	29. +256	30. -343
31. +5	32. +8	33. 0	34. -9
35. -7	36. -17	37. +10	38. -16
39. -16	40. -15	41. 5 qt	42. -3000

43. 26 ft	44. -10 s	45. 396°C	46. 14,778 ft
47. -4	48. -0.2%		
49. $440	50. $7200; $2645		

51. -7 m/s   52. 1650 lb/in.^{2}

53. 1.04 A	54. -7°C	55. -0.8°F	56. 31,184 bytes
57. 5980	58. -518	59. -2	60. -4
61. -5	62. 4	63. 2	64. -9.432

EXERCISES 4-1

1. ab	2. xyz	3. x^2	4. a^4
5. $2w^2$	6. $6a^2c$	7. a^3b^2	8. $6a^2bc^3$
9. b, c	10. 2, a, x	11. 7, p, q, r	12. 3, x, y, z
13. i, i, R	14. 17, a, a, b	15. a, b, c, c, c	16. π, r, r, h
17. 6	18. 3	19. 2π	20. $8a^2$
21. $4\pi e$	22. qBL	23. mw^2	24. $36c$
25. $x = 4y$	26. $x = y^2$	27. $m = 10c$	28. $A = 6e^2$
29. $V = 7.48lwd$			30. $H = Ri^2$

31. $d = \frac{1}{2}gt^2$ 32. $R = 0.00002T^4$

33. $N = 5280x$	34. $A = s^2$	35. $N = 9xy$	36. $C = 3t$
37. $N = 24n$	38. $N = 8x$	39. $C = 4cs$	40. $A = Ns^2$
41. 1500 mi	42. $120	43. 1.5 V	44. 80 cm^3
45. 720 gal	46. 48 J	47. 81.5 ft	48. 9.38 m^2
49. 13,200 ft	50. 108 ft^2	51. 3456 bolts	52. 128,000 m^2
53. 6970 gal	54. 33.84 m^2	55. 172 ft	56. 383 J

57. $40.80   58. 160,000,000 bits

59. 4536 m²   60. 1,074,000 cm³

EXERCISES 4-2

1. x^2, $4xy$, $-7x$

2. a, $2ab$, $-\dfrac{a}{b}$

3. 12, $-5xy$, $7x$, $-\dfrac{x}{8}$

4. $3(x + y)$, $-6a$, $3x$

5. $3x$ and $2x$

6. $-2b$ and $5b$

7. x and $5x$

8. a and $2a$

9. $-8\ mn$ and $-mn$

10. $-bR$ and $-5bR$

11. $6(x - y)$ and $-3(x - y)$

12. $5(a - b)$ and $(a - b)$

13. $6a(a - x)$

14. $3x^2(a - x^2)$

15. $x^2(a - x)(a + x)$

16. $3a^2(a^2 + x^2)(a^2 - x^2)$

17. $\dfrac{2}{5a}$

18. $\dfrac{x^2}{a}$

19. $\dfrac{6}{a-b}$

20. $\dfrac{3x^2 - 2x + 5}{x + 2}$

21. $K = 328$

22. $P = \$880$

23. $t = 2$

24. $C = 20^{\circ}$

25. $L = 500 - x$

26. $p = 2l + 2w$

27. $A = 2x^2 + 4lx$

28. $A = p + prt$

29. $V = I(R + r)$

30. $m = \dfrac{1}{2}(s + l)$

31. $A = 2\pi r(r + h)$

32. $V = \dfrac{p}{t+1}$

33. $A = \dfrac{a+b+c+d+e}{5}$

34. $E = \dfrac{I-P}{I}$

35. $T = 6.28\sqrt{l/g}$

36. $c = \sqrt{a^2 + b^2}$

37. 22.6 ft

38. 134.2 cm

39. 77.8 ft^2 40. 0.357 V   41. 7   42. $1000

43. 2.54 s   44. 0.65   45. 9.741 m^2 46. $1040

47. 3.121 V   48. 18.60 V   49. 2.24 s   50. 0.7015

51. 5.349 m²   52. 2.650 m

EXERCISES 4-3

1. $6x + y$ 2. $2x + 5xy$ 3. $7a - 4b^2$ 4. $7x - 2y$

5. $5s + 3t$ 6. $2m + 2n$ 7. $5x + 5y$ 8. $6s + 6t$

9. $6a - 3b$ 10. $7x - 21y$ 11. $\frac{7x}{2}$ 12. $\frac{4ac}{5}$

13. $\frac{1}{2xy}$ 14. $\frac{1}{2b^2}$ 15. $7a - 15b$ 16. $-2x + 8y$

17. $12x + y$ 18. $12a - 5ab + 4b$

19. $9R + 12S + 5RS$ 20. $18x^2y - 30xy^2z + 12xyz$

21. $-5x + 5y - 4xy$ 22. $-ab - 5bc - 4ac$

23. $4x$ 24. $-3y$ 25. $5rs - 15r$ 26. $-3a + 2ab$

27. $5abc - c^2 - b$ 28. $5mx^2 - 6mx$ 29. $5ax - 3$ 30. $10a^2 - 9$

31. $5x - 10y$ 32. $5bx + 2cx$

33. $8x - 2y + b$ 34. $2a^2 - 4ab + 5bc$

35. $5x^2 + 6xy$ 36. $6a - 7xy$

37. $4x + 5a$ ft   38. $9x - 1000$ papers

39. $20a + 10b + 5$ cents   40. $70t + 80$ kilometers

41. $2xy + 5y + 15$ 42. $180 - 7x - 3y$

43. $3ay$ minutes   44. $25(x - y)$ ohms

45. 40   46. 324   47. 78   48. 420

49. $1.75   50. 220 km   51. 456 cm^2 52. 150 Ω

53. 846   54. 1664   55. -576   56. 196

57. 9921.45   58. 128.6   59. -9.907   60. 39.00

51

EXERCISES 4-4

1. $-7a$ and $5a$ 2. $3ax^2$ and $-6ax^2$

3. $5(a - b)$ and $-7(a - b)$ 4. $8x^2$ and $-2x^2$

5. $7a - 4b$ 6. $3a^2b + 5ab$

7. $12ax - 18bx$ 8. $-24a + 15\ ab$

9. $3x + y$ 10. $x^2 + 6xy$

11. $c - ac$ 12. $4xy - y^2$

13. $2a - b$ 14. $4a^2 + ax$

15. $3a + 11b - ab$ 16. $2x^2 + xy + y$

17. $2ax + 2a$ 18. $2y - 7$

19. $2a - ax$ 20. $4a^2 + 2ac$

21. $5x - xy$ 22. $2ac - 2a$

23. $3ab + a - 2b$ 24. $7x - 4y$

25. $3 + 5a + ax$ 26. $-2x$

27. $13a + 3b$ 28. $3cx - c$

29. $5a + 4b$ 30. $5ac^2 - bc + ab$

31. $4ax - 5xy + 2ay$ 32. a^2

33. $6a^2$ 34. $3ab$

35. $12x$ 36. $15p$

37. 42 38. 27 39. 98 40. 1275

41. -26 42. 1311 43. $218,448$ 44. -192

45. $x = x^2 - 7$ 46. $x + y = 2xy$ 47. $C = \dfrac{Q}{V}$ 48. $F = \dfrac{9}{5}C + 32$

49. $p = 3s$ 50. $S = x^2 + xy$ 51. $I = 1000t^2$ 52. $E = mc^2$

53. $I = 4xr + 0.06x$ 54. $T = 30w - 100$

55. $C = 40n + 145$ 56. $L = \dfrac{a-b}{1000}$

57. $N = 33xt + 20x + 48$ 58. $A = 6xt - 5x$

59. $I = 14 - 4t$ 60. $C = 2ct - 7t$

61. 21 in.   62. 5580 ft   63. \$12,000   64. \$6300

65. (a) \$4.65   (b) \$6.25   66. $129,136$

67. $2,312,736$ 68. (a) \$127   (b) \$358

69. 52 70. $1,227,000$ 71. 190.09 72. -1734

EXERCISES 5-1

1. 2	2. 11	3. 3	4. 9
5. 7	6. 7	7. 35	8. 24
9. -18	10. -40	11. -4	12. -6
13. 9	14. -5	15. -2	16. 6
17. 3	18. 2	19. -6	20. 1
21. -1	22. 5	23. 8	24. 9
25. 1	26. 10	27. 1/2	28. 2
29. $\frac{1}{3}$	30. -4	31. 2	32. $-\frac{9}{2}$
33. 9	34. 8	35. 32	36. 13
37. 3	38. 11	39. 68	40. 3
41. -2	42. 16	43. $\frac{25}{21}$	44. 3

45. First has no solution; second is an identity.
46. First has no solution; second has solution of $x = 2$.
47. a, d 48. a, d, e 49. 0.906 50. -2.27 51. 1.45 52. -1.64

EXERCISES 5-2

1. $\dfrac{N}{A-s}$

2. $\dfrac{D}{2R} + P$

3. $\dfrac{R_1 L_2}{L_1}$

4. $A - AS$

5. $v_2 - at$

6. $C - bx$

7. $\dfrac{PV}{R}$

8. $\dfrac{E}{R}$

9. $\dfrac{Id^2}{5300E}$

10. $\dfrac{2E}{v^2}$

11. $\dfrac{yd}{ml}$

12. $D_0 P - 2$

13. $180 - A - C$

14. $\dfrac{T_d - 3T_2}{-3}$

15. $\dfrac{RM}{CV}$

16. $rL + g_1$

17. $\dfrac{MD_m}{D_p}$

18. $\dfrac{5F - 160}{9}$

19. $\dfrac{L - 3.14r_1 - 2d}{3.14}$

20. $\dfrac{Q_1 + PQ_1}{P}$

21. $\dfrac{L - L_0}{L_0 t}$

22. $\dfrac{a + PV^2}{V}$

23. $\dfrac{p - p_a + dgy_1}{dg}$

24. $\dfrac{F - A_2 + A_1 + PV_1}{P}$

25. $\dfrac{n_1 A + n_2 A - n_2 p_2}{n}$

26. $\dfrac{Qd + kATt_1}{kAT}$

27. $\dfrac{f_s u - fu}{f}$ 28. $\dfrac{V_1(V_2 - V_1)}{gP}$ 29. $a - bc$ 30. $3t + 2y$

31. $\dfrac{f - 3y}{a}$ 32. $\dfrac{7q - b}{3a}$ 33. $\dfrac{3y - 2ay}{2a}$ 34. $\dfrac{2 + 3b}{3}$

35. $2b + 4$ 36. $2x - ax$ 37. $\dfrac{x_1 - x_2 - 3a}{a}$ 38. $\dfrac{s_1 + s_2 + 2b}{2}$

39. $2R_3 - R_1$ 40. $1 - 3m$ 41. $\dfrac{3y + 6 - 7ay}{7a}$ 42. $\dfrac{6 - 2x - 3x^2}{3x}$

43. $\dfrac{x - 3a - ax}{a}$ 44. $\dfrac{a + 2x}{3}$ 45. $3A - a - b$ 46. $\dfrac{T - 400}{3}$

47. $\dfrac{E - Ir}{I}$ 48. $\dfrac{y + 100}{40}$ 49. $\dfrac{I - xr_1}{x + 1000}$ 50. $\dfrac{N - 100t}{2t + 8}$

51. $\dfrac{C - x}{7}$ 52. $\dfrac{K - At}{t - 2}$

EXERCISES 5-3

1. (a) $8 > 6$; (b) $-5 < 1$ 2. (a) $2 < 12$; (b) $3 > -3$
3. (a) $3 > -6$; (b) $1 < 2$ 4. (a) $12 < 40$; (b) $-1 > -2$
5. (a) $2 > 0$; (b) $-2 > -3$ 6. (a) $-6 > -27$; (b) $-4 < 4$
7. (a) $15 > 10$; (b) $48 > -32$ 8. (a) $18 > 12$; (b) $108 > 72$
9. $-1 > -6$ 10. $-1 < 1$ 11. $-1 \geq -2$ 12. $-9 \leq -2$
13. $x > 12$ 14. $x < -4$ 15. $x < 5$ 16. $x > 11$
17. $x > 18$ 18. $x < 20$ 19. $x < -6$ 20. $x \geq 2$
21. $x \leq 4$ 22. $x \geq -5$ 23. $x \leq -1$ 24. $x \geq -2$
25. $x < -4$ 26. $x > 2$ 27. $x < -2$ 28. $x > 3$
29. $x > -1$ 30. $x < -4$ 31. $x \geq 6$ 32. $x \leq 1$
33. Less than \$87,500.   34. Less than or equal to 58.

35. More than 50 mi/h.   36. $\dfrac{80 + 92 + 86 + 78 + 2x}{6} < 80$; $x < 72$

37. $12.9 \leq x \leq 15.5$ 38. $62.6 \leq P \leq 67.2$
39. Greater than 35 h.   40. Greater than 40.
41. $x \leq 2.33$ 42. $x < 2.034$
43. $x \geq 14.69$ 44. $x < 457.28$

EXERCISES 5-4

1. 220 Ω   2. 72, 150   3. $1.20   4. 16 ft
5. 85 mA   6. 9 ft
7. 16 gal/min, 12 gal/min, 22 gal/min   8. 150 m by 200 m
9. 320 lines/min, 800 lines/min   10. 2048 bytes; 16,384 bytes
11. 13 at 5¢, 8 at 25¢   12. 3
13. $21   14. 2500 at $100; 22,000 at $25
15. 800 gal, 1200 gal, 2400 gal   16. 332 mm
17. 4 h   18. 3 h   19. 3600 ft/s   20. 2 qt
21. 75 km/h, 85 km/h   22. 400 mL
23. 16, 64   24. 12.5 lb at 80% nickel and 37.5 lb at 40% nickel.
25. 150 mA, 200 mA, 300 mA   26. 30 t
27. $1800, $2500   28. 150,000 weekday; 200,000 Sunday
29. 1,900,000 first year.; 2,600,000 second year
30. $4000 federal; $800 state   31. $166\frac{2}{3}$ g   32. $2\frac{2}{7}$ qt
33. 50 acres at $200; 90 acres at $300   34. 6 L
35. 48 mi/h   36. 1200 mi   37. 344 m   38. $48,000
39. $11,000   40. $7000, $8000

EXERCISES 5-5

1. $\frac{12}{5}$; $\frac{3}{23}$ 2. $\frac{2}{11}$; $\frac{13}{7}$ 3. $\frac{7}{1}$; $\frac{1}{6}$ 4. $\frac{1}{9}$; $\frac{4}{1}$

5. $\frac{2}{3}$; $\frac{2}{3}$ 6. $\frac{5}{6}$; $\frac{1}{6}$ 7. $\frac{2}{11}$; $\frac{2}{7}$ 8. $\frac{10}{1}$; $\frac{6}{13}$

9. $\frac{15}{4}$ 10. $\frac{2}{11}$ 11. $\frac{3}{10}$ 12. $\frac{25}{8}$

13. $\frac{1}{6}$ 14. $\frac{5}{4}$ 15. $\frac{4}{9}$ 16. $\frac{1}{12}$

17. 2 m/s   18. $\frac{1}{50}$ in./C°   19. $\frac{2}{9}$ lb/ft³   20. $\frac{5}{7}$ L/h

21. $\frac{5}{4}$ 22. $\frac{7}{4}$ 23. $\frac{6}{7}$ 24. $\frac{24}{5}$

25. 5   26. 96   27. 9   28. $\frac{96}{5}$

29. 14

30. $\frac{17}{4}$

31. $\frac{35}{2}$

32. $\frac{32}{3}$

33. $\frac{1}{4}$

34. $\frac{15}{1}$

35. 4540 g

36. 45 mi/h

37. $\frac{100}{3}$ in.

38. $\frac{120}{7}$ lb

39. 6 m, 9 m   40. $2000 federal, $600 state
41. 4,000,000 gal reg; 10,000,000 gal lead-free
42. 180 mg, 100 mg   43. 1.44 m   44. 2.7
45. 1.2   46. 0.78
47. $21,700   48. 0.68108 g/cm^3
49. 6 h   50. 3   51. 37 h   52. 179 gal
53. 9.3 mm   54. 210 m   55. $\frac{9}{5}$ 56. 9 Ω

EXERCISES 5-6

1. $y = kt$

2. $x = ks$

3. $y = ks^2$

4. $s = kt^3$

5. $t = \frac{k}{y}$

6. $y = \frac{k}{x^2}$

7. $y = kst$

8. $s = kxyz$

9. $y = \frac{ks}{t}$

10. $y = \frac{ks}{t^2}$

11. $x = \frac{kyz}{t^2}$

12. $v = \frac{ks^3}{t}$

13. $y = 5s$

14. $y = \frac{14}{t}$

15. $s = 2t^3$

16. $v = 3st$

17. $u = \frac{272}{d^2}$

18. $q = \frac{15}{\sqrt{p}}$

19. $y = \frac{9x}{t}$

20. $t = \frac{35n}{p}$

21. 16

22. 90

23. $\frac{32}{5}$

24. $\frac{15}{2}$

25. 36

26. 36

27. $\frac{36}{49}$

28. 1

29. $\frac{72}{5}$

30. 2700

31. 81

32. 0.5

33. $v = 32t$

34. $E = 23I$

35. $p = 30{,}000t$

36. $P = \frac{RT}{V}$

37. $F = \dfrac{kQ_1Q_2}{s^2}$

38. 18.8 footcandles

39. 35.4 hp

40. 7680 Btu/h

41. 0.116 Ω

42. 31,250 kg·m^2/s^2

43. 240 r/min

44. 0.0000128 per degree F

45. $F = 0.005231\ Av^2$

46. $H = 5.00tl^2$

47. 27.79 s

48. 1.4228 m/s^2

EXERCISES 5-7

1. 8

2. -3

3. 9

4. $\dfrac{1}{2}$

5. -4

6. 9

7. 4

8. 3

9. -3

10. -2

11. $\dfrac{1}{2}$

12. $\dfrac{23}{2}$

13. $R - R_1 - R_2$

14. $\dfrac{wa}{F}$

15. $\dfrac{ms_1}{r}$

16. $\dfrac{P}{l^2}$

17. $\dfrac{d + A}{A}$

18. $\dfrac{qT_2 - wT_2}{q}$

19. $\dfrac{PT - M_2V_2}{V_1}$

20. $\dfrac{f_s u - fu}{f}$

21. $\dfrac{wL}{R(w+L)}$

22. $\dfrac{-\mu R_0 - AR_0}{A}$

23. $\dfrac{W + H_1 - H_2}{S_1 - S_2}$

24. $\dfrac{2p + dv^2 + 2dw}{2d}$

25. $\dfrac{2a + ax}{3}$

26. $\dfrac{bx + bc - c^2}{cx}$

27. $\dfrac{ax + ab - bx}{b}$

28. $\dfrac{2a - 3}{a}$

29. $\dfrac{6 - x - a^2}{a}$

30. $-\dfrac{2r_1r_2}{3}$

31. $\dfrac{2 - 3a}{6}$

32. $\dfrac{2x + 30}{x}$

33. $x < 9$

34. $x < -2$

35. $x \geq 5$

36. $x \geq 3$

37. $x < -3$

38. $x < 4$

39. $x < -4$

40. $x < 1$

41. $x < -2$

42. $x < -4$

43. $x \leq 3$

44. $x \leq 6$

45. $\dfrac{2}{3}$

46. $\dfrac{8}{15}$

47. $\dfrac{6}{1}$

48. $\dfrac{1}{4}$

49. $\frac{8}{3}$ 50. $\frac{17}{3}$ 51. $\frac{9}{2}$ 52. 12

53. 4   54. 8   55. 4   56. $-\frac{5}{2}$

57. $\frac{7}{3}$ 58. $\frac{25}{4}$ 59. $\frac{21}{2}$ 60. $\frac{65}{3}$

61. $y = 6x$ 62. $s = 15t^2$ 63. $m = \frac{15}{\sqrt{r}}$ 64. $v = \frac{24}{\frac{3}{z}}$

65. 5   66. $\frac{243}{8}$ 67. $\frac{81}{5}$ 68. $\frac{216}{5}$

69. 36 in., 54 in.   70. 15 m
71. $150, $50   72. 3.2 ppm, 0.8 ppm
73. 37.2 ft by 24.8 ft   74. 95
75. 48 ft   76. $10,700; $7700; $3200
77. 480 m   78. 9 mA; 3 mA
79. 8 at 14¢; 13 at 22¢   80. 87 dimes, 53 quarters
81. 2.5 h   82. 550 mi/h
83. 1850 km/h, 2150 km/h   84. 4.5 mi
85. 10 L   86. 3.6 mL   87. $9000   88. $19,500
89. 20 lb   90. 27¢
91. 3.51 s   92. $T_2 = \dfrac{Ht + kAT_1}{kA}$

93. 2.45   94. $x > -0.2025$
95. 71.5820 mi/h   96. $F = 167.8x$

EXERCISES 6-1

1. $26°$ 2. $65°$ 3. $104°$ 4. $246°$

5. $56°24'$ 6. $18°54'$ 7. $136°27'$ 8. $79°3'$

9. $156.25°$ 10. $33.8°$ 11. $67.1°$ 12. $16.95°$

13. $\angle CBD, \angle DBA$ 14. $\angle ABC$ 15. $\triangle CDB$ 16. $\triangle ABD$

17. $\triangle ADC$ 18. 54

19. $AC = 4$ in., $BC = 3$ in.   20. $AB = 5$ in.

21. AB∥DC, AD∥BC

22. ∠A and ∠C; ∠B and ∠D

23. 135°   24. 8 cm

25. AB and DC   26. 3 in.

27. AO or OB   28. 4 cm

29.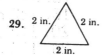

30.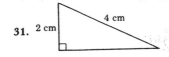

31.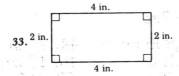

32.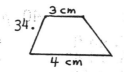

33.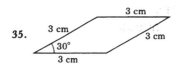

34. 3 cm / 4 cm trapezoid

35.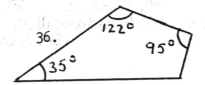

36. 122° 95° 35° quadrilateral

37. (a) Yes; (b) No

38. (a) Yes; (b) No

39. Two isosceles triangles.

40. rhombus

41. 10 cm   42. 115°   43. 54°   44. $2\frac{5}{8}$ in.

45. 50°   46. 10°   47. 94 km   48. 800 ft

49. 66°   50.   51. West   52. South

53. 21°   54. 37.511111° 55. 46°3'36''   56. 13°9'18''

59

EXERCISES 6-2

1. 12 ft	2. 55 cm	3. 123 mm	4. 61 in.
5. 896 m	6. 165 yd	7. 84.8 in.	8. 33.18 cm
9. 193.8 mm	10. 384 ft	11. 57.1 in.	12. 84.9 cm
13. 2.6 m	14. 9.44 yd	15. 0.60 mi	16. 712 mm
17. 168.0 cm	18. 6.30 in.	19. 230.6 ft	20. 11.08 m
21. 94.8 cm	22. 66.6 ft	23. 21.2 in.	24. 132 mm
25. 36 ft	26. 25.10 mm	27. 23.00 cm	28. 12 ft 8 in.
29. 45.36 in.	30. 12.49 cm	31. 24.22 m	32. 285.82 ft

33. $p = 2s + a$ 34. $p = 4s$

35. $p = 2r + \pi r$ 　　　　36. $p = 2r + \frac{1}{2}\pi r$

37. $p = 5s$ 　　　　38. $p = a + b + r + \frac{1}{2}\pi r$

39. $p = 2a + b_1 + b_2$ 　　40. $p = 8a + 2\pi a$

41. $33	42. 45 ft	43. 24,900 mi	44. 21 in.
45. 890 ft	46. $4.40	47. $1312	48. 12 cm
49. 25.1 ft	50. 2680 m	51. 1428 ft	52. 106 yd
53. 26.00 in.	54. 12.15 in.	55. 1.58 mm	56. 8.89 cm
57. 6.28 in.	58. 18.85 cm	59. 4.58 cm	60. 245 cm

61. 234.3 cm　62. First three digits agree.

63. Answer varies. 　　　　64. Answer varies.

EXERCISES 6-3

1. 2700 cm^2	2. 12,900 ft^2	3. 170 in.^{2}	4. 13.3 m^2
5. 57.8 in.^{2}	6. 0.0256 km^2	7. 6.45 in.^{2}	8. 4.0 cm^2
9. 2450 mm^2	10. 1.80 yd^2	11. 24.4 cm^2	12. 252 ft^2
13. 106 in.^{2}	14. 36,000 cm^2	15. 40 ft^2	16. 74.6 m^2
17. 0.240 m^2	18. 464 in.^{2}	19. 340 cm^2	20. 57.8 ft^2
21. 238 ft^2	22. 49,500 mm^2	23. 15,700 cm^2	24. 821 in.^{2}
25. 130 cm^2	26. 4.35 m^2	27. 12 ft^2	28. 64.3 in.^{2}
29. 0.000258 km^2			30. 3.75 yd^2

31. 31.0 m^2 32. 3.27 ft^2 33. 0.717 ft^2 34. 22.7 cm^2

35. 1692 ft^2 36. 6890 m^2

37. 21.65 in.^{2} 38. 32,300 mm^2 39. 1.27 ft^2 40. 12.00 m^2

41. (a) 36 in.; (b) 60 in.^{2} 42. (a) 24.0 m; (b) 32 m^2

43. (a) 60 cm; (b) 225 cm^2 44. (a) 194.6 ft; (b) 1710 ft^2

45. (a) 30 in.; (b) 30 in.^{2} 46. (a) 41.4 mm; (b) 137 mm^2

47. (a) 26.6 in.; (b) 35.0 in.^{2} 48. (a) 12.03 yd; (b) 8.60 yd^2

49. $A = bh_1 + \frac{1}{2}bh_2$ 50. $A = 2ab - a^2$

51. $A = r^2 + 2rh + \frac{1}{2}\pi r^2$ 52. $A = \frac{5}{2}r^2 - \frac{1}{4}\pi r^2$

53. $179.20   54. 176 ft^2 55. 30 ft^2 56. 1.9 gal

57. 10.5 ft^2 58. 7.26 cm^2 59. 122,000 m^2 60. 213,000 cm^2

61. 1070 cm^2 62. 9   63. 196 cm^2 64. $12,900

65. Multiplied by 4.   66. Multiplies by 9.

67. circle   68. 25; 78.5%   69. 46,010,000 ft^2 70. 865.9 m^2

71. 0.4729 m^2 72. 3.658 m^2

EXERCISES 6-4

1. 20,100 mm^3 2. 104 ft^3 3. 0.080 cm^3 4. 45 cm^3

5. 4600 m^3 6. 1600 in.^{3} 7. 1,500,000 cm^3 8. 10,500 ft^3

9. 3400 cm^3 10. 1700 ft^3 11. 0.512 in.^{3} 12. 820 m^3

13. 0.011 m^3 14. 5180 in.^{3} 15. 7156 cm^3 16. 5259 cm^3

17. 85,200 cm^3 18. 2,000,000 ft^3

19. 14,700 m^3 20. 708,000 mm^3

21. 21,000 in.^{3} 22. 233,000,000 cm^3

23. 0.735 mm^3 24. 290 km^3 25. 1728 in.^{3} 26. 46,656 in.^{3}

61

27. $1,000,000 \text{ cm}^3$ 28. $1,000,000,000 \text{ mm}^3$

29. 1300 ft^3 30. 410 cm^3 31. 90 ft^3 32. $23,000 \text{ in.}^3$

33. 15.9 34. 3010 gal 35. No   36. 290 yd^3

37. 4.71 m^3 38. 450 in.^3 39. 311 in.^3 40. 10.3 cm

41. 11.9 lb 42. $465,000 \text{ mm}^3$

43. $1,390,000,000,000,000 \text{ km}^3$ 44. Multiplied by 8.

45. $21,200 \text{ gal}$ 46. 626 yd^3 47. 1.271 mm^3 48. 0.721 ft

EXERCISES 6-5

1. $37°30'$ 2. $43°57'$ 3. $12°33'$ 4. $45°16'12''$

5. $63.5°$ 6. $82.33°$ 7. $105.9°$ 8. $215.75°$

9. 40.2 in. 10. 158.1 cm 11. 25.5 mm 12. 1.29 yd

13. 27.2 m 14. 60.8 in. 15. 1798 ft 16. 28.0 m

17. 278 cm 18. 400 in. 19. 26.7 ft 20. 119 cm

21. 153 mm 22. 1.68 mi 23. 15.8 mm 24. 25.6 ft

25. 224 cm^2 26. 8840 in.^2 27. 3.08 yd^2 28. $15,000 \text{ mm}^2$

29. 4.80 cm^2 30. 1460 ft^2 31. 3.06 in.^2 32. 398 m^2

33. 0.00148 km^2 34. 432 in.^2

35. 35.4 in.^2 36. $20,700 \text{ cm}^2$

37. $140,000 \text{ mm}^2$ 38. 0.290 ft^2

39. 3.30 m^2 40. 9.47 ft^2 41. 1200 mm^2 42. 0.14 mi^2

43. 9.30 mm^2 44. 31.7 ft^2 45. 3.60 m^3 46. 3750 in.^3

47. 42.9 yd^3 48. $10,600,000 \text{ mm}^3$

49. 678 cm^3 50. 125 ft^3 51. 27.4 yd^3 52. 4.032 m^3

53. $p = 2b + 2c$ 54. $p = 5a$

55. $p = 3a + \pi a$ 56. $p = 2r + 3\pi r$ 57. $A = \frac{1}{2}ac$ 58. $A = \frac{3}{2}ah$

59. $A = \frac{1}{2}\pi a^2 + \frac{3}{2}ah$ 60. $A = \frac{3}{2}\pi r^2$

61. $19,440   62. 140 ft, 208 ft

63. 204 m^2 64. $967   65. 30,700 lb   66. 18

67. 15 cm   68. 508.0 mm   69. 26,200 mi   70. 128,000 lb

71. 6.30 ft^2 72. 8400 ft^2; 13,000 ft^2

73. 8.77 ft^2 74. 137 lb   75. 47.6 m^3 76. 275 in.^{3}

77. 9.26 cm^3 78. 18.6 lb   79. 1.77 m^3 80. 116 cm^3

81. 150 gal   82. 50 acres

83. 63,000 ft^3 84. 24 in.

85. 54 in.^{3} 86. 14,000,000 Btu

87. 1.41 ft^2 88. 17,600 rev   89. 49.4 in.^{2} 90. 217 ft

91. 28,300 cm^3 92. 53.6 in.^{2} 93. 10,200 ft^3 94. 37°13'48"

95. 51.396111° 96. 82,500 cm^2

EXERCISES 7-1

1. $3s - xy - a$
2. $-6t^2 - 9as - p + 2h$
3. $9y - 3x - 4a - 9xy$
4. $6u - 4rs - 5y - 9s$
5. $3x + 2xy$
6. $-2as - 11py - 5s$
7. $-3x^2 + 3xy - 2s$
8. $-13y + 5w - u + 15uy$
9. $2a + 3$
10. $x + 4$
11. $-4 - 4x$
12. $5y + 8$
13. $2s - 1$
14. $2x + 3y + 2$
15. $-9 + 4y$
16. $-3b^2 + 3as + 2x^2 - 2s$
17. $10x + 1$
18. $6n + 1$
19. $2s + 2$
20. $4x^2 + 3$
21. $6x - 12$
22. $b + 4 - 7x$
23. $7t - 5x - 5p^2 + 9$
24. $34 - 9xy$
25. $2a^2 + 3x + 1$
26. $6s + 12 + 8x$
27. $7 + 2t$
28. $a^2 - 3xy$
29. 1
30. $-\dfrac{3}{2}$

31. $\dfrac{3}{2}$ 32. $\dfrac{13}{10}$ 33. $5 - 4a$ 34. $\dfrac{4b - c}{6a}$

35. $\dfrac{6}{5}$ 36. $\dfrac{2a - 2}{3}$ 37. $2x + 28$ 38. $x + 3000$

39. $2R + 30$ ohms 40. $2x - 100$ dollars

41. $2M - 40$ 42. $v_1 - v_2$

43. $16t_1^2 - 16t_2^2 + 32$ 44. $B_2 + C_2 - B_1 - C_1$

45. $\dfrac{I + mv}{m}$ 46. $2V - 2198.6$

47. $C + P_1 - L$ 48. 4 ft by 8 ft

49. -0.2840 50. -0.058

51. 1.21 52. 2.8

EXERCISES 7-2

1. x^{10} 2. n^8 3. y^8 4. p^{10}

5. x^3 6. $\dfrac{1}{x^6}$ 7. a 8. y^7

9. t^{10} 10. x^{24} 11. n^{14} 12. t^{15}

13. $\dfrac{1}{p^{10}}$ 14. $\dfrac{1}{p^8}$ 15. $4n^3$ 16. $\dfrac{3}{2m^4}$

17. y^{15} 18. r^{32} 19. p^{25} 20. x^{36}

21. $-ax$ 22. $-x^2 y^2$ 23. $\dfrac{2at^4}{c^2}$ 24. $-\dfrac{3c^2 p^3}{4y^5}$

25. a^{40} 26. n^{18} 27. r^{150} 28. t^{14}

29. $-\dfrac{x}{4r^2}$ 30. $\dfrac{t^2}{s^3}$ 31. $\dfrac{7t^2 u}{6}$ 32. $\dfrac{3as}{5d^3}$

33. $a^2x^4b^2$ 34. $-a^9b^3$ 35. $-a^5t^{10}$ 36. $c^6x^{12}y^6$

37. $-\dfrac{x}{3y^2}$ 38. $\dfrac{2b}{3c}$ 39. $\dfrac{t^3}{3s^3}$ 40. $-\dfrac{1}{2s^{10}}$

41. $-84r^2s^2t^4$ 42. $14a^3x^5y^8$ 43. $-8s^3t^9x^3$ 44. $81a^4x^4t^{28}$

45. True: e only   46. True: b, c
47. True: a, b, c   48. True: b, d, e
49. $2x^3$ 50. n^{28} 51. $I = \dfrac{E}{R}$ 52. $\dfrac{4cr}{3}$

EXERCISES 7-3

1. $2a^2 + 6ax$

2. $6x^2 - 15x$

3. $6a^2x - 3a^4$

4. $12b^3 + 8b^5$

5. $-2s^2tx + 2st^3y$

6. $-64y^2 - 8t^2y$

7. $-3x^3y^2 + 9ax^2y^7$

8. $5uy^{13} + 5hpy^7$

9. $x^2 - 4x + 3$

10. $t^2 + 7t + 10$

11. $s^2 + s - 6$

12. $x^2 - 10x + 24$

13. $2x^2 + x - 1$

14. $15a^2 + 2a - 8$

15. $10v^2 + 17v + 3$

16. $12v^2 - 17v + 6$

17. $a^2 - 3ax + 2x^2$

18. $x^2 + xy - 2y^2$

19. $6a^2 + ac - 2c^2$

20. $15s^2 + 26st + 8t^2$

21. $2x^2 + 18x - 5tx - 45t$

22. $8x + 12uxy - 18uy - 27u^2y^2$

23. $4a^2 - 81p^2y^2$

24. $-8s^2 + 23su^2x + 3x^2u^4$

25. $2a^3 - 5a^2 - 13a - 5$ 26. $6x^3 - x^2 - 11x + 6$

27. $a^2 + 2axy - 4ax - 2x^2y + 3x^2$ 28. $x^3 - 3x^2 - 3x + 10$

29. $x^2 - 4x + 4$ 30. $a^2 + 10a + 25$

31. $x^2 + 4xy + 4y^2$ 32. $4a^2 - 12ab + 9b^2$

33. $x^3 - 3x^2 - 10x + 24$ 34. $6x^3 - 11x^2 - 5x + 12$

35. $x^3 + 3x^2 + 3x + 1$ 36. $8a^3 - 12a^2x + 6ax^2 - x^3$
37. True: c only   38. True: b only
39. True: b, c   40. True: a only
41. $x^2 + x - 6$ feet   42. $320x - 88x^2 + 6x^3$

43. $200x^2 + 1600x + 3000$ 44. $(-5)(-7) = -8 - 27$

45. $(13)(7) = 100 - 9$ 46. $4w - wx^2 - 4wx + wx^3$

47. $16t^2 - 64t + 48$ 48. $nr_2 - nr_1 - r_2 + r_1$

49. $kL_1T_1 - kL_1T_2 - kL_2T_1 + kL_2T_2$

50. $nV_0p_1 - nV_0p_2 - nV_jp_1 + nV_j p_2$

51. $8\,\Omega,\ 3\,\Omega$ 52. $P + 2Pr + Pr^2$

EXERCISES 7-4

1. $6b + 5$ 2. $2x - 3$ 3. $3m - 1$ 4. $4s^2 + s$

5. $-a^2x^2 + ax$ 6. $-y^3 - 2xy^2$ 7. $y^2 - 2xy^3$ 8. $2pq^3 - 7p^2$

9. $bc - ab^3c^4 - 2a$ 10. $-4st^3 + 9r^2t^2 + 8t$

11. $-ab^2 + 2a^2b^3 + 1 + b$ 12. $-my + 6 - 7mn^6$

13. $x - 3$ 14. $x + 5$

15. $2x + 1$

16. $3x - 2 - \dfrac{1}{x + 2}$

17. $4x - 3 + \dfrac{2}{2x + 3}$

18. $2x + 3$

19. $2x^2 - 3x - 4$

20. $2x^2 - 6x + 5 - \dfrac{37}{2x + 5}$

21. $2x^2 - x - 3$

22. $3x^2 - 2x + 1$

23. $x^3 + x^2 + x + 1$

24. $x^2 + 2x + 4$

25. $5(x - 1)$

26. $x^5 - x^2 - 6x - 17 - \dfrac{54}{x - 3}$

27. $2x^3 - x^2 + 2x + 1$

28. $3x^3 - 3x^2 - 2x - 4 + \dfrac{49}{2x + 7}$

29. $a - 3b$

30. $2x - 3y$

31. $x^2 - 4xy - 2y^2$

32. $2s^2 - 3st + t^2$

33. True: b, c

34. True: c only

35. True: a, b

36. True: a, c

37. $a + 5$

38. $c - 8$

39. $a^2 - 5$

40. $b^2 + 1$

41. $2x + 3$ kg

42. $\dfrac{1}{R_1} + \dfrac{1}{R_2} + \dfrac{1}{R_3}$

43. $6r - 4 + \dfrac{18}{r + 2}$

44. $x + 2$

45. $5x - 6 + \dfrac{12}{x + 2}$

46. $3x - 14$ mi/h

47. $10r - 2 + \dfrac{2.4}{6r + 1.2}$

48. $4x + 16$

49. $2.01x^4 - 0.691x^3 + 2.98$

50. $1.428x^6 - 1.133x^2 + 2.669$

51. $21.61x^2 - 2.643x - 81.73$

52. $4.107x^3 + 0.6202x - 41.28$

EXERCISES 7-5

1. $3a - x$ 2. $3x - 3s$ 3. $9x - 13y$ 4. $-7s$

5. $2n - 4$ 6. -12 7. $8y$ 8. $6r - 2x$

9. $-6a^3b^6$ 10. $8s^4t^6$ 11. $56x^4y^9z^8$ 12. $-12x^5y^7$

13. $27a^3b^6$ 14. $16a^{16}c^4$ 15. $16x^8y^4z^{12}$ 16. $-x^{15}y^{10}z^{20}$

17. $-4ax^3$ 18. $\dfrac{3s^2}{4r^3t}$ 19. $\dfrac{5x}{y^4z^3}$ 20. $\dfrac{8b^4}{3a^6c^4}$

21. $8 - 5x$ 22. $2y - 2$

23. $5x - 2a$ 24. $-4x - 5xy + 3y$

25. $4x - 7$ 26. $9a - 6$ 27. $5b - 10$ 28. $4x - 5y$

29. $2x^5 - 6x^3$ 30. $3s^7 - 2s^4$

31. $-2a^3x + 2a^2t$ 32. $-9a^2j^5 + 12a^3j - 3a^3j^2$

33. $2x^2 + x - 21$ 34. $3x^2 + 13x - 10$

35. $6a^2 - 11ab - 10b^2$ 36. $-10x^2 + 7xy - y^2$

37. $x^3 + 1$ 38. $2x^3 - 5x^2 - 5x + 6$

39. $-2x^3 + 6x^2 + 4x - 16$

40. $-3x^2y + 6xy^2 - 9qxy + 3qx - 6qy + 9q^2$

41. $-2y^3 + 3x^3$ 42. $3a^2b^2 - 4b^3$

43. $h - 3j^2 - 6h^3j^3$ 44. $3f^2g - 4fk^4 + 6k^2$

45. $x + 4$ 46. $3x - 5$

47. $x^2 - x + 1 - \dfrac{11}{2x + 3}$ 48. $2x^2 - 3x + 1 + \dfrac{2}{3x + 5}$

49. $x^2 - x + 1$ 50. $2x^2 + x + 1$

68

51. $2x^2 - x - 4$ 52. $3x^2 + 2x + 2$ 53. $x + 2y$ 54. $3a^2 + 2b$

55. $x^2 + x + 1$ 56. $x - 3$ 57. 7   58. $5b - 1$

59. $a^2 + 1$ 60. $2c - 3$

61. $1.71C + 33.6$ 62. $2T_1 - 2T_2$

63. $\dfrac{1 - C - 3I_t}{3}$ 64. $V - iR - ir + E$

65. $-4x^2 + 8x$ 66. $x - 11$ units   67. $36x - 4x^2$ 68. $80t - 32t^2$

69. $mv_2^2 - mv_1^2$ 70. $\dfrac{ID}{d}$

71. $lh + 2alht + a^2lht^2$ 72. $8dx$

73. $r_1 - ar_1 + r_2 - 2ar_2 + a^2 r_2$ 74. $100x - 40x^2 + 4x^3$

75. $p_0 - kx$ 76. $1 - \dfrac{Q_2}{Q_1}$

77. $2a + 5$ 78. $a + ar + ar^2 + ar^3$

79. $2x^2 + 20x + 50$; 450 lb   80. 372 mi

81. $2841x^8$ 82. $6.68x^2 - 3.20x + 3.96$

83. $1024x^{10} - 256x^8 - 1$ 84. 2.068

EXERCISES 8-1

1. $5(x + y)$ 2. $3(x^2 - y)$ 3. $7(a^2 - 2bc)$ 4. $3(s - 4t)$

5. $a(a + 2)$ 6. $p(3 + q)$ 7. $2x(x - 2)$ 8. $5h(1 + 2h)$

9. $3(ab - c)$ 10. $4x(x - 1)$ 11. $2p(2 - 3q)$ 12. $3s(2s^2 - 5t)$

13. $3y^2(1 - 3z)$ 14. $12x(1 - 4y)$ 15. $abx(1 - xy)$ 16. $2xy(1 - 4xy)$

17. $6x(1 - 3y)$ 18. $4xy(3 - 2a)$ 19. $3ab(a + 3)$ 20. $3xy^2(2y - 3)$

21. $acf(abc - 4)$ 22. $2rst(s - 4rt)$ 23. $ax^2y^2(x + y)$ 24. $2a^2x^3(y - 3)$

25. $2(x + y - z)$ 26. $3(r - s + t)$

27. $5(x^2 + 3xy - 4y^3)$ 28. $2(2rs - 7s^2 - 8r^2)$

29. $2x(3x + 2y - 4)$ 30. $5s(s^2 + 2s - 4)$

31. $4pq(3q - 2 - 7q^2)$ 32. $6x^2y(3y - 4y^2 + 9x)$

33. $7a^2b^2(5ab^2c^2 + 2b^3c^3 - 3a)$ 34. $x^2yz(15z^2 - 45xyz + 16y)$

35. $3a(2ab - 1 + 3b^2 - 4ab^2)$ 36. $4r^2s(1 - 2rs + 4r^2 - s^2)$

37. $nR\left[\dfrac{T_2}{T_1} - 1\right]$ 38. $2\pi r(r + h)$

39. $I(R_1 + R_2 + R_3)$ 40. $t(I_1^2R_1 + I_2^2R_2)$

41. $2wh(7 + 2w)$ 42. $2x^2(4x^2 - 3x + 5)$

43. $2(lw + lh + hw)$ 44. $2t(225 - 8t)$

45. $2x(x^2 - 3x + 5)$ 46. $2iMC(A + B - 5D)$

47. $wx(x^3 - 2Lx^2 + L^3)$

48. $P = Rv(1 + v + v^2 + v^3 + v^4 + v^5)$

49. (a) Yes; (b) No, $2x(x - 4)$ 50. (a) No, $6a(x^2 - 2)$; (b) Yes

51. (a) Yes; (b) No   52. (a) Yes; (b) Yes

53. (a) Yes; (b) No; (c) No   54. (a) Yes; (b) No; (c) No

55. (a) No; (b) Yes; (c) No   56. (a) No; (b) Yes; (c) No

EXERCISES 8-2

1. $\sqrt{9} = 3$; $\sqrt{16} = 4$ 2. $\sqrt{4x^2} = 2x$

3. $\sqrt{121} = 11$ 4. $\sqrt{x^4} = x^2$; $\sqrt{9x^4} = 3x^2$

5. $\sqrt{x^6} = x^3$; $\sqrt{16x^6} = 4x^3$

6. $\sqrt{x^8} = x^4$; $\sqrt{144x^8} = 12x^4$; $\sqrt{144a^2x^8} = 12ax^4$

7. $\sqrt{a^2b^4} = ab^2$ 8. $\sqrt{4r^6} = 2r^3$ 9. $x^2 - y^2$ 10. $a^2 - 4$

11. $4a^2 - b^2$ 12. $9x^2y^2 - 1$ 13. $49a^2x^4 - p^6$ 14. $25x^4 - 36y^{10}$

15. $x^2 + 4x + 4$ 16. $a^2 + 2ab + b^2$ 17. $(a + 1)(a - 1)$ 18. $(x+2)(x-2)$

19. $(t + 3)(t - 3)$ 20. $(E + 4)(E - 4)$

21. $(4 + x)(4 - x)$ 22. $(5 + y)(5 - y)$

23. $(2x^2 + y)(2x^2 - y)$ 24. $(5s^3 + 1)(5s^3 - 1)$

25. $(10 + a^2b)(10 - a^2b)$ 26. $(8 + xy^2)(8 - xy^2)$

27. $(ab + y)(ab - y)$ 28. $(2q + rs)(2q - rs)$

29. $(9x^2 + 2y^3)(9x^2 - 2y^3)$ 30. $(7b^3 + 5c^2)(7b^3 - 5c^2)$

31. $5(x^2 + 3)(x^2 - 3)$ 32. $6(2x + 3a)(2x - 3a)$

33. $4(x + 5y)(x - 5y)$ 34. $9(x + 3)(x - 3)$

35. $(x^2 + 1)(x + 1)(x - 1)$ 36. $(a^2 + 9b^2)(a + 3b)(a - 3b)$

37. $4(x^2 + 9y^2)$ 38. $3(25a^2x^2 + 9b^4y^4)$

39. $3s(t^2 + 4)$ 40. $9a^2b^2(b^4 + 9)$

41. $8(2x^2y^2 + 3a^2b^2)$ 42. $4x^3(3a^2 - 1)$

43. $5ax(x - 8a)$ 44. $7y^6(x^2 + 5)$

45. $(x + y + 1)(x + y - 1)$ 46. $(a - b + 3)(a - b - 3)$

47. $(5 + x + y)(5 - x - y)$ 48. $(6 + s + t)(6 - s - t)$

49. $4xy$ 50. $4x$

51. $a(x + y + 1)(x + y - 1)$ 52. $ax^2(x + y + 3)(x + y - 3)$

53. 399 54. 6396 55. $39,900$ 56. $22,464$

57. $(c + 8)(c - 8)$ 58. $b(a + c)(a - c)$

59. $c\pi(r_1 + r_2)(r_1 - r_2)$ 60. $(L + 2x)(L - 2x)$

61. $(C + kp)(C - kp)$ 62. $m(v_1 + v_2)(v_1 - v_2)$

63. $kw(h_2 + h_1)(h_2 - h_1)$

64. $\pi h(r_1 + r_2)(r_1 - r_2)$

65. $k(T_2^2 + 1)(T_2 + 1)(T_2 - 1)$

66. $a^2 b^2 c^2 (1 + ab)(1 - ab)$

67. $2\pi d(h_1 a_1 + h_2 a_2)(h_1 a_1 - h_2 a_2)$

68. $x(x + y)(x - y)$

EXERCISES 8-3

1. $(x + 1)(x + 2)$
2. $(x - 2)(x + 1)$
3. $(x + 4)(x - 3)$
4. $(s + 3)(s + 4)$
5. $(y - 5)(y + 1)$
6. $(x + 7)(x - 1)$
7. $(x + 5)^2$
8. $(t + 5)(t - 2)$
9. $(2q + 1)(q + 5)$
10. $(2a - 3)(a + 1)$
11. $(3x + 1)(x - 3)$
12. $(2x - 5)(x - 1)$
13. $(5c - 1)(c + 7)$
14. $(3x + 7)(x - 1)$
15. Prime
16. Prime
17. $(2s - 3t)(s - 5t)$
18. $(3x - 2)(x - 4)$
19. $(5x + 2)(x + 3)$
20. $(3x + 1)(x - 6)$
21. $(2x - 3)(2x - 1)$
22. $(3x - 1)(2x + 7)$
23. $(6q + 1)(2q + 3)$
24. $(8y - 3)(y + 1)$
25. $(6t - 5u)(t + 2u)$
26. $(4x + y)(x + 8y)$
27. $(4x - 3)(2x + 3)$
28. $(4x + 3y)(3x - 4y)$
29. $(4x - 3)(x + 6)$
30. $(3s + 4)(2s - 5)$
31. $(4n - 5)(2n + 3)$
32. $(9y - 8)(y + 4)$
33. $2(x - 3)(x - 8)$
34. $3(x - 5)(x + 1)$
35. $2(2x - 3z)(x + 2z)$
36. $5(x^2 + 3x + 5)$
37. $2x(x + 1)(x + 2)$
38. $x^2(x - 2)(2x + 5)$
39. $a(5x - y)(2x + 5y)$
40. $3(6a - 13b)(3a + 4b)$
41. $3a(x + 5)(x - 3)$
42. $r^2(2x + 3)(x - 5)$
43. $7a^3(2x + 1)(x - 1)$
44. $5c(x^2 + x + 3)$
45. $(x + 1)^2$ 46. $(x - 3)^2$
47. $(x - 4)^2$ 48. $(x + 6)^2$
49. $(2x + 1)^2$ 50. $(2x + 3)^2$
51. $(3x - 1)^2$ 52. $(4x - 3)^2$
53. $N(r + 1)^2$
54. $(x - L)(x - 2L)$
55. $2(p - 4)(p - 50)$
56. $P(R + 1)^2$
57. $2(x - 4)(x - 8)$
58. $5(T + 20)(T + 100)$
59. $4x(x - 6)^2$
60. $k(3x + 1)(2x + 3)$
61. $3x(2x - 1)(x - 2)$
62. $A(kD + 1)^2$
63. $4(4t + 1)(t - 8)$
64. $2(x - 40)(x - 50)$

EXERCISES 8-4

1. $8 = 2^3$
2. $27 = 3^3;\ 8x^3 = (2x)^3$
3. $64 = 4^3;\ 27x^3 = (3x)^3$
4. $125 = 5^3;\ 64x^3 = (4x)^3$
5. $x^6 = (x^2)^3;\ 8x^6 = (2x^2)^3$
6. $125x^9 = (5x^3)^3;\ 8a^3x^6 = (2ax^2)^3$
7. $a^3x^9 = (ax^3)^3$
8. $125x^6y^9 = (5x^2y^3)^3$
9. $x^3 + y^3$
10. $x^3 - y^3$
11. $x^3 + 8$
12. $x^3 - 27$
13. $8a^3 - b^3$
14. $8a^3 + 27b^3$
15. $x^6 - 8$
16. $x^6 + y^9$
17. $(a - 1)(a^2 + a + 1)$
18. $(a + 1)(a^2 - a + 1)$
19. $(t + 2)(t^2 - 2t + 4)$
20. $(t - 3)(t^2 + 3t + 9)$
21. $(1 - x)(1 + x + x^2)$
22. $(2 + s)(4 - 2s + s^2)$
23. $(2x + 3a)(4x^2 - 6ax + 9a^2)$
24. $(4s - 3t)(16s^2 + 12st + 9t^2)$
25. $(2x^2 - y)(4x^4 + 2x^2y + y^2)$
26. $(x + 3y^3)(x^2 - 3xy^3 + 9y^6)$
27. $(ax - y^2)(a^2x^2 + axy^2 + y^4)$
28. $(ax^2 + b^2y)(a^2x^4 - ab^2x^2y + b^4y^2)$
29. $8x(x + 1)(x^2 - x + 1)$
30. $12xy(1 - y)(1 + y + y^2)$
31. $ax^2(1 - y)(1 + y + y^2)$
32. $4(x^2 + 1)(x^4 - x^2 + 1)$
33. $2k(R_1 + 2R_2)(R_1^2 - 2R_1R_2 + 4R_2^2)$
34. $(I_1R_1 + I_2R_2)(I_1^2 R_1^2 - I_1I_2R_1R_2 + I_2^2 R_2^2)$
35. $2(3t_1^2 - t_2)(9t_1^4 + 3t_1^2 t_2 + t_2^2)$
36. $3(L_1 - 3L_2^4)(L_1^2 + 3L_1L_2^4 + 9L_2^8)$
37. $(5s^2 - 4t^3)(25s^4 + 20s^2t^3 + 16t^6)$
38. $2ax(5x - 4)(25x^2 + 20x + 16)$
39. $(ab + c^5)(a^2b^2 - abc^5 + c^{10})$
40. $(a^3 + bc^4)(a^6 - a^3bc^4 + b^2c^8)$
41. $(1 - ax)(1 + ax)(1 + ax + a^2x^2)(1 - ax + a^2x^2)$
42. $(x^2 - y)(x^2 + y)(x^4 + x^2y + y^2)(x^4 - x^2y + y^4)$
43. $(1 - x - y)(1 + x + y + (x+y)^2)$
44. $(x + y + 1)((x+y)^2 - x - y + 1)$
45. $N(x - y)(x^2 + xy + y^2)$

46. $K(T_1 - T_2)(T_1^2 + T_1T_2 + T_2^2)$

47. $\frac{4}{3}\pi(r_1 + r_2)(r_1^2 - r_1r_2 + r_2^2)$

48. $N(s + r)(s^2 - sr + r^2)$

49. $k(r_1 - r_2)(r_1^2 + r_1r_2 + r_2^2)$

50. $(2x + 3)(4x^2 - 6x + 9)$

51. $k\pi(r_1 - r_2)(r_1^2 + r_1r_2 + r_2^2)$

52. $C(x + y)(x^2 - xy + y^2)$

53. $V(T_1 - T_2)(c_1 - c_2)(c_1^2 + c_1c_2 + c_2^2)$

54. $at^3(t - 1)(t^2 + t + 1)$

55. $\frac{nw^2}{24p^2}(L_1 - L_2)(L_1^2 + L_1L_2 + L_2^2)$

56. $(x - 2y)(x^2 + 2xy + 4y^2)$

EXERCISES 8-5

1. $5(a - c)$
2. $4(r + 2s)$
3. $3a(a + 2)$
4. $2t^2(3t - 4)$

5. $4ab(3a + 1)$
6. $5ty(3t^2 - 2y)$
7. $8stu^2(1 - 3s^2)$
8. $4xyz^4(4x - 1)$

9. $(2x + y)(2x - y)$
10. $(p + 3uv)(p - 3uv)$

11. $(4y^2 + x)(4y^2 - x)$
12. $(rst + 2x)(rst - 2x)$

13. $(x + 1)^2$
14. $(x - 3)(x + 1)$

15. $(x - 1)(x - 6)$
16. $(x + 9)(x - 7)$

17. $a(x^2 + 3ax - a^2)$
18. $3r^2t(6 - 3rt - 2t^2)$

19. $2nm(m^2 - 2nm + 3n^2)$
20. $8y^2(1 + 3yz - 4z^4)$

21. $4t^2(p^3 - 3t^2 - 1 + a)$
22. $11rst^2(2rs - 11 - 2r^2 + 3r^3s)$

23. $2xy^3(1 - 7x + 8y - 3x^2y^2)$ 24. $3st^2(1 - 2s^2tu - 4u + 3t)$

25. $(4rs + 3y)(4rs - 3y)$ 26. $(7r^2t^2 + y^3)(7r^2t^2 - y^3)$

27. Prime   28. $(a + b + c)(a + b - c)$

29. $(2x + 7)(x + 1)$ 30. $(3y - 5)(y + 2)$

31. $(5s + 2)(s - 1)$ 32. $(3a - 1)(a + 7)$

33. $(7t + 1)(2t - 3)$ 34. $(5x - 4)(x + 1)$

35. $(3x + 1)^2$ 36. $(4r - 5)(2r + 3)$

37. $(x + y)(x + 2y)$ 38. $(3a - 4b)(2a - 3b)$

39. $(5c - d)(2c + 5d)$ 40. $(2p - 3q)^2$

41. $(8x + 7)(11x - 12)$ 42. $(4y + 7)^2$

43. $2(x + 3y)(x - 3y)$ 44. $4(rt + 3pq)(rt - 3pq)$

45. $8x^4y^2(xy + 2)(xy - 2)$ 46. $3mn(m^2 + 3n)(m^2 - 3n)$

47. $3a(x - 3)(x + 4)$ 48. $2x(9c + 5)(2c - 3)$

49. $3r(6r - 13s)(3r + 4s)$ 50. $4c^2(2x + 9)(x + 2)$

51. $16y^3(y - 1)(y - 3)$ 52. $a^2(9u - 2)(2u + 3)$

53. $5(x^2 + 5)(x^2 - 5)$ 54. $4(a^4 + 4)(a^2 + 2)(a^2 - 2)$

55. $(4x^2 + 1)(2x + 1)(2x - 1)$ 56. $(x^4 + 1)(x^2 + 1)(x + 1)(x - 1)$

57. $(x + 3)(x^2 - 3x + 9)$ 58. $(t - 4)(t^2 + 4t + 16)$

59. $(2x + 1)(4x^2 - 2x + 1)$ 60. $8(x^2 - y)(x^4 + x^2y + y^2)$

61. $axy(x - y)(x^2 + xy + y^2)$ 62. $ab(b + a)(b^2 - ba + a^2)$

63. $(5R_1 + 2aR_2)(25R_1^2 - 10aR_1R_2 + 4a^2R_2^2)$

64. $a^3(x^4 - ay^3)(x^8 + ax^4y^3 + a^2y^6)$

65. $i(R_1 + R_2 + R_3)$ 66. $16t(4 - t)$

67. $P(N + 2)$ 68. $k(R + r)(R - r)$

69. $k(D + 2r)(D - 2r)$ 70. $P(V_2 - V_1)(C + R)$

71. $(v_2 - 3v_1)(v_2 - v_1)$ 72. $\frac{1}{2}r(v_1 + v_2)(v_1 - v_2)$

73. $\dfrac{(u-1)^2}{(u+1)^2}$ 74. $b(x+y)(x-y)$

75. $(T-10)(T+530)$ 76. $2(x+5)(x-5)$

77. $V(T_1-T_2)(k_1-k_2)(k_1{}^2+k_1k_2+k_2{}^2)$

78. $c(t+1-t^3)(1+2t+t^2+t^3+t^4+t^6)$

79. $\dfrac{4}{3}\pi(2r_1+3r_2)(4r_1{}^2-6r_1r_2+9r_2{}^2)$

80. $(x-y)(x^2+xy+y^2)$

EXERCISES 9-1

1. $\dfrac{4}{6}; \dfrac{10}{15}$ 2. $\dfrac{21}{15}; \dfrac{42}{30}$ 3. $-\dfrac{15}{20}; -\dfrac{-15}{-20}$ 4. $-\dfrac{-10}{-14}; -\dfrac{10}{14}$

5. $\dfrac{6a^3x}{2a^2b}; \dfrac{3a^2bx}{ab^2}$ 6. $\dfrac{10bx}{6x^2}; \dfrac{35b^3x}{21b^2x^2}$ 7. $\dfrac{-10ax^3}{-2bx}; \dfrac{10a^2x^3}{2abx}$ 8. $\dfrac{3a^3x^4}{ab^2x}; \dfrac{-3a^3x^5}{-ab^2x^2}$

9. $\dfrac{3x}{x^2-2x}; \dfrac{3x+6}{x^2-4}$ 10. $\dfrac{21a}{3a^2+9a}; \dfrac{7a+7}{a^2+4a+3}$

11. $\dfrac{x^2-2xy+y^2}{x^2-y^2}; \dfrac{x^2-y^2}{x^2+2xy+y^2}$ 12. $\dfrac{2x^2-8y^2}{2x^2+3xy-2y^2}; \dfrac{4x^2-6xy-4y^2}{4x^2-y^2}$

13. $\dfrac{4}{7}$ 14. $\dfrac{9}{13}$ 15. $\dfrac{2a}{3a^2}$ 16. $\dfrac{2rs}{4st}$

17. $\dfrac{2}{x+1}$ 18. $\dfrac{x-3}{3}$ 19. $\dfrac{2x-1}{x+1}$ 20. $\dfrac{3x-1}{x}$

21. $\dfrac{3(x-2)}{x+2}$ 22. $\dfrac{x-3}{2(x+2)}$ 23. $\dfrac{-(2+x)}{x-3}$ 24. $\dfrac{-2(x-3)}{3+x}$

25. $\dfrac{2}{9}$ 26. $\dfrac{2}{3}$ 27. $\dfrac{1}{3}$ 28. $\dfrac{1}{6}$

29. $3x$ 30. $\dfrac{3a}{2}$ 31. $\dfrac{ab}{4}$ 32. $\dfrac{4}{7z}$

33. $\dfrac{8}{9}$ 34. $\dfrac{4x^4}{5}$ 35. $\dfrac{2t}{7r^2s}$ 36. $\dfrac{7a^2}{2c^4}$

37. $\dfrac{2x-1}{x-2}$ 38. $\dfrac{2x-1}{2x+1}$ 39. $\dfrac{(x+1)(x+2)}{2(x+3)}$ 40. $\dfrac{x+5}{3(5x+1)}$

41. $\dfrac{x+1}{x-1}$ 42. $\dfrac{x+1}{x-1}$ 43. $\dfrac{x}{x+2}$ 44. $\dfrac{(4x-3)(x+3)}{4x(x-2)}$
(Not reducable)

45. $\dfrac{3x-2}{4x+3}$ 46. $\dfrac{2x+3}{2(3x-1)}$ 47. $\dfrac{x+3y}{3y}$ 48. $\dfrac{a+8b}{a+b}$

49. $\dfrac{1-3x}{3x+1}$ 50. $\dfrac{x-2}{2x-1}$ 51. $\dfrac{5-x}{2+x}$ 52. $\dfrac{-b}{b+4a}$

53. $-3x$ 54. $\dfrac{-5a}{2+a}$ 55. $\dfrac{2-x}{3+x}$ 56. $\dfrac{1-x}{4}$

57. $\dfrac{3}{7}$ 58. $\dfrac{11}{8}$ 59. All of them.   60. -1

EXERCISES 9-2

1. $\dfrac{1}{8n}$; $13s$ 2. $\dfrac{1}{a^2b}$; $\dfrac{b}{a^2}$ 3. $\dfrac{3b}{a}$; $\dfrac{a}{3b}$ 4. $-\dfrac{y}{2x^2}$; $-\dfrac{3ax}{5cd}$

5. $\dfrac{x-y}{x+y}$; $\dfrac{x^2}{x^2+y^2}$ 6. $-\dfrac{s-t}{s^2t^2}$; $\dfrac{16-y^2}{x^2-9}$ 7. $\dfrac{a+b}{a}$; $-\dfrac{V}{IR}$ 8. $\dfrac{3}{4\pi r^2}$; $\dfrac{w}{2L+2H}$

9. $\dfrac{4}{9t}$ 10. $\dfrac{2}{3}$ 11. $\dfrac{2a}{15}$ 12. $\dfrac{42x}{13a}$

13. $\dfrac{rt}{12}$ 14. $\dfrac{75xz}{y}$ 15. $\dfrac{81}{256}$ 16. $\dfrac{343}{125}$

17. $\dfrac{a^{10}}{32x^5}$ 18. $\dfrac{x^5y^{20}}{32z^{10}}$ 19. $\dfrac{a^3x^9}{b^6}$ 20. $\dfrac{y^{18}}{64x^6}$

21. $\dfrac{26}{35cx}$ 22. $\dfrac{66ac}{85}$ 23. $\dfrac{24mx}{7}$ 24. $\dfrac{27y}{20z}$

25. $\dfrac{y}{45x}$ 26. $\dfrac{5a^3x^2}{8y^2}$ 27. $\dfrac{1}{2a^2b^5}$ 28. $\dfrac{3}{5x^2}$

29. $\dfrac{a+3b}{a+b}$ 30. $\dfrac{n+2}{9}$ 31. $\dfrac{5x(x-y)}{6}$ 32. $\dfrac{3a}{4(a+b)}$

33. $\dfrac{x+1}{x+2}$ 34. $\dfrac{2a-1}{a}$ 35. $\dfrac{(x+3)(x-3)}{(x-2)(x-4)}$ 36. $\dfrac{3(x+1)}{4(x-1)}$

37. $\dfrac{5b - 2}{10}$ 38. $\dfrac{9x}{x + 1}$ 39. $\dfrac{3(a - b)}{(a-2b)(a+b)}$ 40. $\dfrac{2(x + 3)}{x - 3}$

41. $\dfrac{(s+2)(s+7)}{(s - 12)(s+11)}$ 42. $\dfrac{(p + 2q)(p + q)}{(2p - q)(p - 3q)}$

43. $\dfrac{(x - 1)^2}{3x + 2}$ 44. 1 45. $\dfrac{2a(a + b)}{(a-b)(2a+b)}$ 46. $\dfrac{1}{2x + 5}$

47. $\dfrac{81a(2x - 3y)(2x + 3y)}{(x - y)(x - y)}$ 48. $\dfrac{-(3x + 7)^2}{(4x + 5)(2x - 5)(x + 2)}$

49. $\dfrac{pq}{p + q}$ 50. $\dfrac{mv^2}{6}$ 51. $\dfrac{1}{79f^2c}$ 52. $\dfrac{3.2(x + 2)}{x}$

53. $\dfrac{4000}{x}$ 54. $\dfrac{1 - 2n + n^2}{1 + 2n + n^2}$ 55. $I = \dfrac{2t(t+1)}{3(2t+1)}$

EXERCISES 9-3

1. 18	2. 14	3. 36	4. 120
5. $12a$	6. $75x$	7. $40t$	8. $200R$
9. $90y$	10. $360n$	11. $9x^2$	12. $392t^3$
13. $8x^2$	14. $15a^3$	15. $420ax$	16. $40rst$
17. $375ax^2$	18. $48a^2b^2$	19. $75a^3$	20. $8x^5$
21. $96a^3b^3$	22. $27a^2bc^5$	23. $15a^2$	24. $8x^3$
25. $60a^2cx^3$	26. $675p^2q^2rs^2$	27. $8x(x - 1)$	28. $4x(x + 4)$
29. $3a(a + 3)$	30. $12y^2(y - 2)$	31. $6ax(a - 3)$	32. $6a^2x^2(4x + a)$

33. $6x(x - y)(x + y)$ 34. $4x^3(x - 1)(x + 1)$

35. $(a - 2b)(a + 2b)(a + b)$ 36. $2(x - 1)^2(x - 2)$

37. $12(x - 3)(x + 3)^2$ 38. $2(2t + 3)(t - 4)(t^2 + 5t + 3)$

39. $2(x - 3y)(x + 3y)(3x + 2y)$ 40. $12(x^2 + y^2)(x - y)^2(x + y)$

EXERCISES 9-4

1. $\dfrac{20}{36a} - \dfrac{21}{36a}$

2. $\dfrac{21}{840ax} + \dfrac{50x}{840ax}$

3. $\dfrac{5b}{abx} + \dfrac{a}{abx} - \dfrac{4bx}{abx}$

4. $\dfrac{72}{12a^3b} - \dfrac{15a^2}{12a^3b} - \dfrac{2ab}{12a^3b}$

5. $\dfrac{8x}{2x^2(x-1)} - \dfrac{3}{2x^2(x-1)}$

6. $\dfrac{5b(b+2c)}{3a(b-2c)(b+2c)} - \dfrac{7ac}{3a(b-2c)(b+2c)}$

7. $\dfrac{x(x+2)^2}{2(x-2)(x+2)^2} + \dfrac{10(x+2)}{2(x-2)(x+2)^2} - \dfrac{6x(x-2)}{2(x-2)(x+2)^2}$

8. $\dfrac{3(a-1)(a-4)}{9(a-1)(a-4)} - \dfrac{72}{9(a-1)(a-4)} - \dfrac{(a-1)(a-4)}{9(a-1)(a-4)}$

9. $\dfrac{19}{10x}$

10. $-\dfrac{1}{24s}$

11. $\dfrac{2b-5a}{3ab}$

12. $\dfrac{7t+2s}{8st}$

13. $\dfrac{6x+3}{x^2}$

14. $\dfrac{8a-15}{6a^2}$

15. $\dfrac{26b-25}{40b}$

16. $\dfrac{149s-32}{120s}$

17. $\dfrac{8y-b}{by^2}$

18. $\dfrac{a+2}{a^2x}$

19. $\dfrac{9+2x^2}{3x^3y}$

20. $\dfrac{30-p}{6p^2q}$

21. $\dfrac{2xy+5x^2-3}{x^2y}$

22. $\dfrac{48b-20a^2+9a}{24a^2b}$

23. $\dfrac{42yz-15xz+2xy}{12xyz}$

24. $\dfrac{1-13q}{6pq}$

25. $\dfrac{4(2a-1)}{(a-2)(a+2)}$

26. $\dfrac{10x+3}{(2x-1)(2x+3)}$

27. $\dfrac{2x^2+3x+9}{4(x-3)(x+3)}$

28. $\dfrac{a^2b+ab^2+2a-2b}{a^2(a+b)(a-b)}$

29. $\dfrac{y^2-6y-3}{3(y-3)(y+3)}$

30. $\dfrac{2x^2+x+2}{2x(x-4)}$

31. $\dfrac{-(x+6)}{2(2x+3)}$

32. $\dfrac{2c(17d-10c)}{3(5c-2d)(5c+2d)}$

33. $\dfrac{3x^2-17x+14}{(2-3x)(2+3x)}$

34. $\dfrac{2q^2+q+3}{q(q-4)(q+3)}$

35. $\dfrac{1 + 15x - 5x^2}{(x - 2)(x + 2)}$

36. $\dfrac{11x - 7}{2(x - 1)(x + 1)}$

37. $\dfrac{2x^2 + 3x - 125}{3(x + 5)(x - 5)}$

38. $\dfrac{3x^3 + 35x^2 - 2x - 4}{x(3x - 1)(3x + 1)}$

39. $\dfrac{7p^2 + 4p + 8q - 175q^2}{8(p - 5q)^2}$

40. $\dfrac{4(8 - x)}{(x + 3)(x - 3)(x + 5)}$

41. $\dfrac{R_1 + R_2}{R_1 R_2}$

42. $\dfrac{t_1 + t_2 + t_3}{t_1 t_2 t_3}$

43. $\dfrac{41}{24}$ in.^{2}

44. $\dfrac{29}{1280}$ mi^2

45. $\dfrac{120 - 60x^2 + 5x^4 - x^6}{120}$

46. $\dfrac{26.500x^2 + 5.080x^3 + 0.004}{x^2}$

47. $\dfrac{R_2 - R_1}{R_1}$

48. $\dfrac{U - U_d}{U_d U}$

49. $\dfrac{II_0 - I_r - I_t}{I_0}$

50. $\dfrac{rm_1 - 3rm_2}{m_1 + m_2}$

51. $\dfrac{g_m + 8}{g_m{}^2}$

52. $\dfrac{x^2(6w + kx^2)}{12T_0}$

53. $\dfrac{p^2 - 2gm^2rM}{2mr^2}$

54. $\dfrac{au(3\ell + 2u)}{\ell + u}$

55. $\dfrac{h_1 P_1^2 - h_2 P_2^2}{(h_1 + h_2)(h_1 - h_2)}$

56. $\dfrac{n(2n + 1)}{2(n + 2)((n - 1)}$

EXERCISES 9-5

1. 4   2. 2   3. -3 ~ 2   4. 8

5. 10   6. -16   7. $-\dfrac{3}{4}$ 8. $-\dfrac{1}{4}$

9. $1 - 3b$ 10. $3 + 2c$ 11. $4a + 2$ 12. $\dfrac{3b - 1}{3}$

13. $\dfrac{2a - 4}{3a^2}$ 14. $\dfrac{5 - 2c^2}{2c^3}$ 15. $-\dfrac{16}{7b}$ 16. $\dfrac{a^2}{6a - 8}$

17. 2   18. 2   19. $\dfrac{7}{3}$ 20. $-\dfrac{4}{3}$

21. $\dfrac{3}{4}$ 22. $\dfrac{25}{8}$ 23. No solution.   24. No solution

25. $\dfrac{-b}{3b - 1}$ 26. $\dfrac{2a + 7}{3}$ 27. $\dfrac{3}{2n - 4}$ 28. $\dfrac{3p + 14}{2p^2 + 2p}$

29. $\dfrac{52}{11}$ 30. $a = kV - PV^2$ 31. $\dfrac{pf}{p - f}$ 32. $\dfrac{R_1 R_2}{R_1 + R_2}$

33. $\dfrac{2D_p}{D_0 - D_p}$ 34. $\dfrac{WT_1 + Q_1 T_1}{Q_1}$ 35. $\dfrac{2\pi^2 r^2 (P - p)}{m^2}$ 36. $\dfrac{2v^2 mx - 2v^2 y}{m^2 x^2 + x^2}$

37. \$2000   38. 5 in. by 7.5 in.   39. 1.8 mi   40. 2 h

41. 1.7 h   42. 3.3 h   43. 240 h   44. $\dfrac{24}{7}$ h

EXERCISES 9-6

1. $\dfrac{3rt^4}{s^3}$ 2. $\dfrac{-z}{6y^3}$ 3. $\dfrac{a}{3bc^2}$ 4. $\dfrac{4x^2 y}{z^2}$

5. $\dfrac{4}{x - 2y}$ 6. $\dfrac{a}{x + y}$ 7. $\dfrac{p + q}{3 + 2p^2}$ 8. $\dfrac{4a}{5b}$

9. $\dfrac{a}{2b}$ 10. $\dfrac{p + 1}{p + 3}$ 11. $\dfrac{3x + y}{2x - y}$ 12. $\dfrac{2(y - 5)}{3 - 2y}$

13. 18   14. $10x^2$ 15. $12t^2$ 16. abt^3

17. $48b^2 t$ 18. $aby^2 z^4$

19. $20x^2 (x - 2)$ 20. $(x + 5)(x - 3)(x + 3)$

21. $\dfrac{10a}{3x^2}$ 22. $\dfrac{b^2 c^4}{4}$ 23. $\dfrac{15y}{4x}$ 24. $\dfrac{q}{3p^2}$

25. $\dfrac{6}{a}$ 26. $\dfrac{b^2}{ac^2}$ 27. $\dfrac{2bu}{av}$ 28. $\dfrac{40m^3}{81}$

29. $\dfrac{10b - 3a}{5a^2b}$

30. $\dfrac{9x^2 - 10y^2}{12xy}$

31. $\dfrac{10cd + c - 6}{2c^2d}$

32. $\dfrac{3x^2 + 18x - 2a}{12x^3}$

33. $\dfrac{8}{27}$

34. $\dfrac{l^3}{R^3t^3}$

35. $\dfrac{a^4x^4}{81y^8}$

36. $\dfrac{32x^5}{z^{15}}$

37. $\dfrac{2}{x(x + 1)}$

38. $\dfrac{1}{2a(a - 3)}$

39. $\dfrac{x - 5}{4}$

40. $\dfrac{3(x - 3)(x - 4)(x + 3)}{2(x - 1)(x - 5)}$

41. $a(a - 1)$

42. $\dfrac{4x}{x - 1}$

43. $(3x + 2y)(y - 2x)$

44. $\dfrac{3r + s}{4(r - 2s)}$

45. $\dfrac{5x + 9}{x^2(x + 3)}$

46. $\dfrac{8 - 3a}{2a(a - 2)}$

47. $\dfrac{(x - 1).(x - 3)}{(x - 2)^2}$

48. $\dfrac{3x - 6y - 7}{(3x+y)(x-2y)}$

49. $\dfrac{-(2x + 3)^2}{(2x + 5)(x - 3)(x + 3)}$

50. $\dfrac{3x^2 - 2x}{2(x + 4)(x - 2)}$

51. $\dfrac{-2x^3 + 9x^2 - 43x + 15}{x(x + 5)(x - 5)(2x - 1)}$

52. $\dfrac{25x + 32}{8(x + 2)(x - 2)}$

53. $\dfrac{2}{5}$

54. 1

55. $\dfrac{(x + 1)(x - 3)}{(x - 2)(x + 3)}$

56. $\dfrac{3s + 2}{s^2(s + 1)}$

57. 9

58. $\dfrac{34}{63}$

59. $\dfrac{a + 6}{4(b - a)}$

60. $\dfrac{5by}{4b - 5}$

61. No solution

62. No solution

63. $-\dfrac{8}{21}$

64. 6

65. $\dfrac{nle^2E}{2mv}$

66. $\dfrac{8lu}{\pi a^4}$

67. $\dfrac{3r - h}{12r^3}$

68. $\dfrac{24dL^2s - ds^3}{24L^3}$

69. $\dfrac{g_2^2 - 2g_1g_2 + g_1^2}{g_2^2}$

70. $\dfrac{Z_2 - Z_1}{Z_1 + Z_2}$

71. $\dfrac{C^2L^2\omega^4 - 2LC\omega^2 + 1}{C^2\omega^2}$

72. $\dfrac{4c^4d + 2c^2dv^2 + 3dv^4}{2c^5}$

73. $\dfrac{\mu R}{r + \mu R + R}$

74. $\dfrac{1}{i}$

75. $Prt + P$

76. $\dfrac{k(h - U)}{hu}$

77. $\dfrac{q_2 D - df}{d + D}$

78. $\dfrac{CC_2 + CC_3 - C_2C_3}{C_2 - C}$

79. $\dfrac{2akmM}{amv^2 + kmM}$

80. $\dfrac{PP_B(1 - x)}{P_B - P_x}$

81. 3.4 min

82. 3.1 h

83. $150,000

84. 30 min

EXERCISES 10-1

1. $\dfrac{1}{t^5}$

2. $\dfrac{1}{R^3}$

3. $\dfrac{1}{x^4}$

4. $\dfrac{1}{s^8}$

5. x^3

6. a^7

7. $R_1^{\,3}$

8. t^8

9. $\dfrac{3}{c^2}$

10. $\dfrac{1}{3c}$

11. $\dfrac{c}{3}$

12. $3c$

13. 1

14. 1

15. 1

16. 1

17. 5

18. 3

19. $9y^2$

20. $\dfrac{2}{3x}$

21. 3^6

22. $\dfrac{1}{7^8}$

23. 6

24. $\dfrac{1}{9}$

25. $\dfrac{1}{ax}$

26. $\dfrac{b^3}{c^3}$

27. $\dfrac{2}{c^8}$

28. $6x$

29. $\dfrac{y^2}{x^6}$

30. $\dfrac{a^2}{9}$

31. $\dfrac{x^2}{125}$

32. $\dfrac{4}{a^4}$

33. $\dfrac{s}{t^2}$ 34. $\dfrac{n^4}{m^3}$ 35. $\dfrac{x^4}{8y^4}$ 36. $\dfrac{ab^3}{36}$

37. $\dfrac{b^7}{9a}$ 38. $\dfrac{q^8}{27p^2}$ 39. $\dfrac{4}{25a^2b^2}$ 40. $\dfrac{z^2}{x^4y^2}$

41. $\dfrac{y^5}{x^4}$ 42. $\dfrac{s^7t}{r^3}$ 43. $\dfrac{a^5c^2}{18}$ 44. $\dfrac{y^5}{a^8x^5}$

45. $\dfrac{1}{R_1} + \dfrac{1}{R_2}$ 46. $\dfrac{1}{F} = \dfrac{1}{u} + \dfrac{1}{v}$ 47. $N = \dfrac{N_0}{e^{kt}}$ 48. $p = \dfrac{10}{3^x}$

49. $\dfrac{R}{hR + 1}$ 50. $\dfrac{e^{iax}}{e^{i\omega t}}$ 51. $g \cdot cm^{-3};\ m \cdot s^{-2}$ 52. $\dfrac{2\ m^3}{10^3\ c}$

53. $362\ Btu/h \cdot ft^2$ 54. $10^{-1},\ 10^{-3}$

55. $\dfrac{rR}{R + r}$ 56. -4

EXERCISES 10-2

1. 5   2. -2   3. -4   4. 6
5. 4,000,000   6. 3,800,000,000
7. 0.08   8. 0.0000000000703
9. 2.17   10. 0.793   11. 0.00365   12. 8040
13. 3×10^3 14. 4.2×10^5 15. 7.6×10^{-2} 16. 2.9×10^{-3}
17. 7.04×10^{-1} 18. 1.08×10^{-2} 19. 9.21×10^0 20. 1.03×10^1
21. 5.3×10^{-5} 22. 1.006×10^6 23. 2.01×10^9 24. 4.923×10^{-4}
25. 1.55×10^8 26. 4.37×10^7 27. 9.30×10^{-3} 28. 6.79×10^2
29. 4.65×10^{-5} 30. 2.80×10^2 31. 6.740×10^{-9} 32. 4.26×10^1
33. 1.30×10^1 34. 6.86×10^7 35. 1.26×10^7 36. 1.40×10^2
37. 6.5×10^6 g   38. 3,800,000,000,000 m
39. 4.5×10^{-2} m 40. 0.00065 L
41. 3.92×10^{-2} L 42. 0.00000000000115 s

43. 8.06×10^{-5} s   44. 0.0000000000000291 g

45. 2×10^9 Hz   46. 1.75×10^4 mi/h

47. 9.1×10^{-28} g   48. 6.5×10^{-7} m

49. 4,000,000,000,000,000,000,000,000,000 lb

50. 50,000 lb   51. 0.000001 in.   52. 0.0000000015 s

53. 3.6×10^8 km^2 54. 5×10^{-5} F   55. 0.0000000000016 W

56. 1,000,000,000,000,000,000,000,000,000^{o}C

57. 3.6×10^7 mi   58. 7.1×10^8 years   59. 0.000000000001 W/cm^2

60. 0.075 Pa   61. 6×10^{-19} J   62. 1×10^{-9} m

63. 30,000,000,000 cm/s   64. 1,050,000 bytes

65. 2.59×10^{10} cm^2 66. 2.81×10^{-21} m^3 67. 6.06×10^7 Hz

68. 195.2 m   69. 4.90×10^{19} 70. 1.74×10^{-14}

71. 6.9×10^{10} 72. 2.563×10^{-8}

EXERCISES 10-3

1. 7	2. 6	3. -12	4. -9
5. 0.4	6. 0.1	7. -0.2	8. -0.6
9. 20	10. 30	11. -40	12. -60
13. 2	14. 3	15. -2	16. -3
17. -5	18. 5	19. 0.5	20. -0.1
21. 2	22. -5	23. 3	24. 3
25. 2	26. -2	27. $2j$	28. $7j$
29. $-20j$	30. $-30j$	31. $0.7j$	32. $-0.01j$
33. -25	34. -5	35. -20	36. -49
37. -12	38. -64	39. 64	40. 8

41. Rational irrational, rational, rational , imaginary
42. Rational, irrational, irrational, rational, imaginary

43. Irrational, rational, imaginary, irrational, rational
44. Rational, irrational, rational, imaginary, rational
45. 6.00 ft   46. 144 ft²   47. 4.71 s   48. $\frac{9}{5}$

49. $\frac{4}{3}$ 50. 30,000   51. 0.1   52. 3.0000 mm

EXERCISES 10-4

1. $\frac{\sqrt{2}}{2}$ 2. $\frac{\sqrt{3}}{3}$ 3. $\frac{2\sqrt{5}}{5}$ 4. $\frac{\sqrt{6}}{2}$

5. $\frac{\sqrt{a}}{a}$ 6. $\frac{a\sqrt{b}}{b}$ 7. $\frac{\sqrt{ab}}{b}$ 8. $\frac{\sqrt{2rs}}{2s}$

9. $\frac{\sqrt{15}}{5}$ 10. $\frac{\sqrt{14}}{7}$ 11. $\frac{a\sqrt{3a}}{3}$ 12. $\frac{a^2\sqrt{bc}}{bc}$

13. $2\sqrt{3}$ 14. $3\sqrt{3}$ 15. $2\sqrt{7}$ 16. $2\sqrt{11}$

17. $3\sqrt{5}$ 18. $3\sqrt{11}$ 19. $5\sqrt{6}$ 20. $7\sqrt{2}$

21. $7\sqrt{3}$ 22. $9\sqrt{2}$ 23. $9\sqrt{3}$ 24. $8\sqrt{10}$

25. $c\sqrt{a}$ 26. $a^2\sqrt{3}$ 27. $ab\sqrt{a}$ 28. $2a^2\sqrt{3a}$

29. $2ac\sqrt{bc}$ 30. $3a\sqrt{ab}$ 31. $4x^2z^2\sqrt{5yz}$ 32. $4y^3z^3\sqrt{15xy}$

33. $\frac{b\sqrt{3a}}{6}$ 34. $\frac{ce^2\sqrt{11ce}}{22}$ 35. $\frac{x\sqrt{10y}}{5a^4}$ 36. $\frac{y^2\sqrt{130axy}}{20a^2}$

37. $3\sqrt[3]{2}$ 38. $2\sqrt[3]{3}$ 39. $2a\sqrt[3]{a}$ 40. $a^3\sqrt[3]{6a}$

41. $2a^2\sqrt[4]{a}$ 42. $3ab\sqrt[4]{b}$ 43. $3a^2\sqrt[4]{3a^3}$ 44. $2x\sqrt[5]{2x^2}$

45. $3a^2x^3\sqrt[4]{2a^2}$ 46. $3rs^2t^2\sqrt[5]{rt^2}$ 47. $2rs^2t^2\sqrt[7]{2t^2}$ 48. $ab^2\sqrt[8]{ac^5}$

49. $\frac{\sqrt{\sqrt{gl}}}{gl}$ 50. $\frac{\sqrt{LrtM}}{M}$ 51. $\frac{2\sqrt{3.14A}}{3.14}$ 52. $f = \frac{\sqrt{LC}}{2\pi LC}$

53. 80 ft, 160 ft   54. 2304 in.³

55. 107.5, 112.5   56. $x = 0.04$ s²; 141 ft/s

86

57. $V = \dfrac{k\sqrt[3]{PW^2}}{W}$ 58. $r = \dfrac{\sqrt[3]{48\pi^2 V}}{4\pi}$ 59. $d = \dfrac{k\sqrt[3]{16JC^2}}{C}$

60. $r = \dfrac{\sqrt[4]{8\pi^3 nLRD^3}}{\pi D}$

EXERCISES 10-5

1. $2\sqrt{7}$ 2. $2\sqrt{3}$ 3. $4\sqrt{7} + \sqrt{5}$ 4. $12\sqrt{11} - \sqrt{17}$

5. $5\sqrt{3}$ 6. $3\sqrt{5} + 5\sqrt{3}$ 7. $9\sqrt{10}$ 8. $4\sqrt{11}$

9. 10 10. $3\sqrt{3}$ 11. $5\sqrt{3}$ 12. $4\sqrt{3}$

13. $9\sqrt{2}$ 14. $9\sqrt{3}$ 15. $\sqrt{7}$ 16. $-\sqrt{6}$

17. $-\sqrt{2} - 4\sqrt{3}$ 18. $4\sqrt{5} + \sqrt{11}$

19. $4\sqrt{a}$ 20. $3\sqrt{5x}$ 21. $(3 + 4a)\sqrt{2a}$ 22. $(3 + 4a)\sqrt{ac}$

23. $7a\sqrt{2} - 2\sqrt{3a}$ 24. $(x - y)\sqrt{yz} + 2xy\sqrt{z}$

25. $\sqrt{21} - 9\sqrt{2}$ 26. $10 - 6\sqrt{15}$

27. 26 28. $2\sqrt{42} - 7\sqrt{2}$

29. $a\sqrt{b} + 3a\sqrt{c}$ 30. $4\sqrt{a} + 2a^2\sqrt{3}$

31. $1 + \sqrt{6}$ 32. $11 - 5\sqrt{35}$

33. $-17 - 3\sqrt{15}$ 34. $6\sqrt{22} + 22\sqrt{2} - 12 - 4\sqrt{11}$

35. $2a - \sqrt{ac} - 15c$ 36. $(3 - b)\sqrt{2} - \sqrt{b}$

37. $7 + 4\sqrt{3}$ 38. $13 - 4\sqrt{3}$

39. $\dfrac{\sqrt{6} + 2}{2}$ 40. $\dfrac{3 + \sqrt{15}}{3}$ 41. 7 42. $\sqrt{15} - 2$

43. $\dfrac{7 - \sqrt{21}}{4}$ 44. $\dfrac{5 + \sqrt{55}}{-3}$ 45. $\dfrac{4 + 3\sqrt{2}}{2}$ 46. $25 - 7\sqrt{14}$

47. $\dfrac{a - 2\sqrt{ab}}{a - 4b}$ 48. $\dfrac{2x + 3\sqrt{xy}}{4x - 9y}$ 49. $\dfrac{R_1\sqrt{R_2} - R_2\sqrt{R_1}}{R_1 - R_2}$

50. $\dfrac{\sqrt{WL_1} - \sqrt{WL_2}}{L_1 - L_2}$

51. $2a$

52. $\dfrac{\sqrt{2g(h_1 - h_2)}}{2g}$

53. $\dfrac{13 - 2\sqrt{30}}{7}$

54. $(\ell + \sqrt{\ell^2 - k^2})^2 - 2\ell(\ell + \sqrt{\ell^2 - k^2}) + k^2 =$
$$\ell^2 + 2\ell\sqrt{\ell^2 - k^2} + (\ell^2 - k^2) - 2\ell^2 - 2\ell\sqrt{\ell^2 - k^2} + k^2 = 0$$

55. $75\sqrt{2}$ ft

56. $W = \dfrac{\sqrt{k(M + 0.23m)}}{M + 0.23m}$

57. 8.46

58. 1.56

59. 0.473

60. 0.130

EXERCISES 10-6

1. $\sqrt{5}$

2. $\sqrt[3]{7}$

3. $\sqrt[4]{a}$

4. $\sqrt[5]{b}$

5. $\sqrt[5]{x^3}$

6. $\sqrt[4]{x^3}$

7. $\sqrt[3]{R^7}$

8. $\sqrt[7]{s^9}$

9. $a^{1/3}$

10. $b^{1/7}$

11. $x^{1/2}$

12. $(ax)^{1/5}$

13. $x^{2/3}$

14. $a^{3/4}$

15. $b^{8/5}$

16. $s^{1/2}$

17. 3

18. 6

19. 4

20. 5

21. 2

22. 3

23. -2

24. -4

25. 8

26. 27

27. 16

28. 9

29. 27

30. 8

31. 4

32. -243

33. $\dfrac{1}{6}$

34. $\dfrac{1}{10}$

35. $\dfrac{1}{2}$

36. $\dfrac{1}{4}$

37. $\dfrac{1}{2}$

38. $-\dfrac{1}{16}$

39. 18

40. 5000

41. 2

42. $\dfrac{1}{3}$

43. 2

44. 5

45. $\dfrac{1}{48}$

46. $\dfrac{1}{3375}$

47. $-\dfrac{1}{14}$

48. $\dfrac{21}{25}$

49. $a^{3/2}$

50. $b^{2/3}$

51. $a^{3/4}b$

52. $xyz^{7/2}$

53. $x^{29/15}$ 54. $xy^{11/6}$ 55. $x^{5/6}$

56. $\dfrac{x^{3/2}z^{5/3}}{y^{1/2}}$ 57. $x^{5/2} + y^{5/2} = k$ 58. $\dfrac{kNia^2}{\sqrt{(a^2 + b^2)^3}}$

59. $x = A^{1/2}$ 60. $e = V^{1/3}$

61. $\dfrac{(c^2 - v^2)^{1/2}}{c}$ 62. $R = \dfrac{\sqrt{(1 + D_1^2)^3}}{D_2}$

63. $r_p = \dfrac{2}{3DK^{2/3}I^{1/3}}$ 64. $E = 100(1 - \dfrac{\sqrt[5]{R^3}}{R})$

65. 4.72×10^{22} 66. $V_1^{1/3} + V_2^{1/3} + V_3^{1/3}$ cm

67. $B = \dfrac{2\pi kI}{R}$ 68. 8.37×10^8 mi

69. 2.1   70. 6.4   71. 6.7   72. 7.6

73. 24.5   74. 1606.1   75. 10.3   76. 0.2

EXERCISES 10-7

1. $\dfrac{1}{10}$ 2. $\dfrac{1}{16}$ 3. 9   4. 4

5. $\dfrac{1}{6}$ 6. 64   7. 13   8. -30

9. 5   10. 5   11. -16   12. 0.14

13. $\dfrac{1}{4}$ 14. $\dfrac{2}{5}$ 15. $-\dfrac{3}{11}$ 16. $-\dfrac{12}{13}$

17. 10   18. 25   19. 10   20. 20

21. 343   22. 27   23. 1331   24. 128

25. $\dfrac{8}{9}$ 26. 625   27. 324   28. $\dfrac{1}{108}$

29. $9j$ 30. $12j$ 31. $-0.8j$ 32. $-0.1j$

33. 5.70×10^2 34. 6.08×10^4 35. 3.25×10^0 36. 1.03×10^{-1}

37. 7.69×10^{-4} 38. 7.75×10^6 39. 8.695×10^1 40. 1.024×10^3

41. 3000 42. 0.0006 43. $314{,}000$ 44. 0.0000675

45. $\dfrac{3b}{a^2}$ 46. $\dfrac{2x}{y}$ 47. $\dfrac{m^4}{n^2}$ 48. $\dfrac{2rt^5}{s}$

49. $\dfrac{2x^5}{3y^3}$ 50. $\dfrac{c^4}{4a^2b^2}$ 51. $-\dfrac{x^2y^3}{a^2}$ 52. $\dfrac{b^2}{3a}$

53. $a^{7/12}$ 54. $x^{13/15}$ 55. $a^{7/6}$ 56. $\dfrac{1}{b^{1/2}}$

57. $\dfrac{x^{1/2}}{y}$ 58. $\dfrac{2y^{1/2}}{x}$ 59. $s^{2/3}t^{7/3}$ 60. $\dfrac{8c^{17/10}}{a}$

61. $2\sqrt{11}$ 62. $3\sqrt{3}$ 63. $6\sqrt{2}$ 64. $3\sqrt{6}$

65. $8\sqrt{2}$ 66. $2\sqrt{31}$ 67. $2\sqrt[3]{5}$ 68. $3\sqrt[3]{4}$

69. $\dfrac{\sqrt{11}}{11}$ 70. $\dfrac{\sqrt{66}}{11}$ 71. $\dfrac{2\sqrt{7}}{7}$ 72. $\dfrac{6\sqrt{37}}{37}$

73. $2a$ 74. $2\sqrt{7a}$ 75. $5b\sqrt{5c}$ 76. $3b\sqrt{10b}$

77. $\dfrac{\sqrt{6a}}{a}$ 78. $\dfrac{\sqrt{3}}{a}$ 79. $\dfrac{2\sqrt{21a}}{3a}$ 80. $\dfrac{200\sqrt{10ab}}{ab}$

81. $2a\sqrt[3]{2}$ 82. $3y\sqrt[3]{3x^2y}$ 83. $2a\sqrt[5]{2a^3}$ 84. $a\sqrt[7]{64a}$

85. $-\sqrt{7}$ 86. $-6\sqrt{5}$ 87. $5\sqrt{7} - 2\sqrt{6}$ 88. $2\sqrt{3} - \sqrt{15}$

89. $\sqrt{70} - 5\sqrt{10}$ 90. $8\sqrt{5} - 7\sqrt{3}$

91. $(6 - 3a)\sqrt{2}$ 92. $(4a - 7)\sqrt{5a}$ 93. $-6\sqrt{3}$ 94. $18\sqrt{2}$

95. $2\sqrt{15} - 30$ 96. $30 + 12\sqrt{3}$ 97. $a\sqrt{b} - 3\sqrt{5ab}$ 98. $b\sqrt{a} - 3a\sqrt{b}$

99. $27 - 5\sqrt{30}$ 100. $-3 - 5\sqrt{14}$ 101. $81 - 17\sqrt{42}$ 102. $27 + 5\sqrt{33}$

103. $2a - 3b - 5\sqrt{ab}$ 104. $2ab - \sqrt{abc} - c$

105. $\dfrac{2 + \sqrt{10}}{3}$ 106. $-\dfrac{3\sqrt{7} + 6\sqrt{3}}{5}$

107. $5\sqrt{2} - 7$ 108. $\dfrac{9 - \sqrt{55}}{13}$ 109. 2.75×10^{10} Hz   110. 1.56×10^7

111. 2×10^{-7} in.   112. 3.15×10^7 s

113. 0.00000000000000000016 C

114. 31,000,000,000,000,000 ft   115. 0.0000000000005 m

116. 0.0000006 kg/m³   117. 9800 J   118. $\dfrac{r^2 + 4R^2}{2\pi R}$

119. $\dfrac{r_1 r_2}{(\mu - 1)(r_2 - r_1)}$ 120. $v = \dfrac{\sqrt{Ed}}{d}$

121. $v = \dfrac{\sqrt{2emV}}{m}$ 122. $\dfrac{Bx^2\sqrt{2qmV}}{4mV}$

123. $\dfrac{21}{8}$ in.   124. $\dfrac{\omega\sqrt{\omega_0^2 + \omega^2}}{\omega_0^2 + \omega^2}$

125. 6.4 V   126. 5.86×10^5 m³   127. 288   128. 0.019

129. $\dfrac{\sqrt{3}}{3}$ 130. $v = \dfrac{331\sqrt{273(273 + T)}}{273}$

131. 0.182 ft   132. 1960 C

133. 7.478×10^{18} 134. 7.74×10^{-10}

135. 1.0045×10^2 136. 9.050×10^2

137. 2.67   138. 10.0   139. 9.37   140. 2.47

141. 28.77   142. 0.0153   143. 0.000842   144. 3.1

EXERCISES 11-1

1. $4x^2 - 3x + 2 = 0$ 2. $x^2 + 9 = 0$ 3. $-x^2 + 8x = 0$

4. $9x^2 + 6x + 12 = 0$ 5. $x^2 = 0$ 6. $2x^2 = 0$

7. $x^2 - x - 1 = 0$ 8. $-3x^2 + 8x - 7 = 0$

9. $x^2 - 7x - 4 = 0$; $a = 1$, $b = -7$, $c = -4$

10. $3x^2 + 9x - 5 = 0$; $a = 3$, $b = 9$, $c = -5$

11. Not quadratic   12. Not quadratic

13. $x^2 + 4x + 4 = 0$; $a = 1$, $b = 4$, $c = 4$

14. $x^2 + x = 0$; $a = 1$, $b = 1$, $c = 0$

15. $7x^2 - x = 0$; $a = 7$, $b = -1$, $c = 0$

16. $4x^2 - 2x = 0$; $a = 4$, $b = -2$, $c = 0$

17. Not quadratic   18. Not quadratic

19. $-6x^2 + 3x - 1 = 0$; $a = -6$, $b = 3$, $c = -1$

20. $x^2 + 6x + 1 = 0$; $a = 1$, $b = 6$, $c = 1$

21. 2, 3   22. 2   23. 2   24. None

25. -1, 2   26. -1, -3   27. $\frac{1}{2}$, 1   28. 1, $-\frac{2}{3}$

29. 3, -4   30. $y^2 - 3y - 10 = 0$; 5 is a solution.

31. $3x^2 - 10x + 8 = 0$; none of listed values

32. $3t^2 + 5t + 2 = 0$; $-\frac{2}{3}$, -1

33. $s^2 - 4s + 4 = 0$; 2   34. -2, 0

35. $n^2 - 9 = 0$; -3, 3   36. None of listed values

37. $-x^2 + 6x - 5 = 0$ 38. $-5x^2 + 8x + 132 = 0$

39. $16t^2 - 96t + 128 = 0$ 40. $605 = 500(1 + r)^2$ becomes
$500r^2 + 1000r - 105 = 0$.

41. $w(w + 2) = 48$ becomes $w^2 + 2w - 48 = 0$.

42. $(-5 + 15j\sqrt{11})^2 + 10(-5 + 15j\sqrt{11}) + 2500 =$
$25 - 150j\sqrt{11} - 2475 - 50 + 150j\sqrt{11} + 2500 = 0$

43. $4\pi r^2 - \pi r^2 = 4$ becomes $3\pi r^2 - 4 = 0$.

44. $3s^2 - 12 = 0$

EXERCISES 11-2

1. 3, -3   2. $\frac{5}{2}$, $-\frac{5}{2}$ 3. -2, 1   4. 2, 3

5. -2, 5   6. -1, 7   7. -2, $\frac{1}{3}$ 8. $\frac{3}{4}$, 1

9. -2, $\frac{1}{2}$ 10. -4, $\frac{2}{3}$ 11. $-\frac{5}{2}$, $\frac{1}{3}$ 12. $-\frac{4}{3}$, $\frac{5}{2}$

13. $-\frac{1}{2}$, $\frac{2}{3}$ 14. $\frac{5}{2}$, 6   15. $\frac{1}{5}$, 4   16. $-\frac{5}{3}$, $-\frac{4}{3}$

17. -1, $\frac{9}{2}$ 18. $-\frac{2}{3}$, $\frac{5}{2}$ 19. -2, -2   20. 3, 3

21. 0, 8   22. 0, $-\frac{5}{3}$ 23. $-\frac{7}{2}$, $\frac{1}{2}$ 24. $-\frac{1}{2}$, $\frac{2}{3}$

25. $\frac{7}{2}$, $\frac{7}{2}$ 26. $-\frac{5}{3}$, $-\frac{5}{3}$ 27. 0, 7   28. 0, 12

29. $-\frac{1}{3}$, 3   30. -1, $\frac{3}{4}$ 31. -2a, 2a   32. 2a, 2a

33. 2, 17   34. 5.556 m/s

35. -6, -2   36. 10 m by 25 m

37. 192°F   38. 2.5 s   39. e = 5 cm   40. x = 2 L, 2 L

41. 1000 mi/h   42. 1, 16

43. 3 ft, 4 ft, 7 ft   44. 5.250 in.

EXERCISES 11-3

1. 3, -1   2. 2, -5   3. $-\frac{1}{2}$, -3   4. 2, $-\frac{1}{3}$

5. $\frac{-5 \pm \sqrt{13}}{2}$ 6. $\frac{3 \pm \sqrt{13}}{2}$ 7. $2 \pm \sqrt{6}$ 8. $1 \pm \sqrt{7}$

9. $\frac{1}{2}$, $\frac{3}{2}$ 10. $-\frac{5}{3}$, 2   11. $\frac{4}{3}$, $-\frac{4}{3}$ 12. 5, -5

13. $-\frac{1}{2}, \frac{7}{2}$

14. $\frac{-7 \pm \sqrt{33}}{4}$

15. $-1 + 2j, -1 - 2j$

16. $\frac{1 \pm j\sqrt{7}}{2}$

17. $\frac{1 \pm \sqrt{33}}{2}$

18. $\frac{5 \pm \sqrt{57}}{2}$

19. $\frac{-1 \pm \sqrt{33}}{4}$

20. $-1 \pm \sqrt{7}$

21. $0, 7$

22. $0, \frac{7}{5}$

23. $\frac{3 \pm \sqrt{55}\, j}{4}$

24. $\frac{9 \pm \sqrt{15}\, j}{8}$

25. $-\frac{2}{3}, 4$

26. $\frac{1}{2}, 7$

27. $-\frac{1}{3a}, -\frac{3}{2a}$

28. $-1, -\frac{a}{2}$

29. $-3, 1$

30. $-7, 1$

31. $-1, -\frac{1}{2}$

32. $\frac{-q \pm \sqrt{q^2 - 4pr}}{2p}$

33. $16, 17$

34. 15

35. 12 cm by 17 cm

36. 3.90 ft, 5.90 ft

37. 8.50 s, 0.441 s

38. $\frac{E \pm \sqrt{E^2 - 4PR}}{2R}$

39. 0.382 atm

40. $m = \frac{-RC \pm \sqrt{R^2C^2 - 4LC}}{2LC}$

41. ± 4 A

42. 3.0 cm

43. 64.9 m

44. 10 in. by 20 in.

45. $-0.34, -1.94$

46. $-1.98, 0.25$

47. $4.20, -0.05$

48. $-0.62, 1.29$

EXERCISES 11-4

1. $-3, -4$

2. $8, -1$

3. $6, -\frac{1}{2}$

4. $-\frac{1}{2}, \frac{3}{2}$

5. $-\frac{1}{6}, 6$

6. $0, \frac{9}{8}$

7. $\frac{3}{4}, \frac{3}{4}$

8. $\frac{7}{2}, \frac{7}{2}$

9. $0, -\dfrac{7}{5}$ 10. $\dfrac{1}{9}, 6$ 11. $-11, 10$ 12. $-20, \dfrac{1}{2}$

13. $3, -8$ 14. $7, -3$ 15. $-3, -3$ 16. $1, \dfrac{3}{2}$

17. $\dfrac{5}{6}, 1$ 18. $-4, \dfrac{4}{7}$ 19. $1, 1$ 20. $0, -8$

21. $-2 \pm \sqrt{2}$ 22. $\dfrac{-1 \pm \sqrt{5}}{2}$ 23. $\dfrac{-5 \pm \sqrt{33}}{4}$ 24. $\dfrac{-2 \pm \sqrt{10}}{3}$

25. $3 \pm \sqrt{15}$ 26. $\dfrac{5 \pm \sqrt{85}}{10}$ 27. $\dfrac{-1 \pm j\sqrt{39}}{4}$ 28. $\dfrac{1 \pm j\sqrt{107}}{6}$

29. $-2j, 2j$ 30. $\pm \dfrac{5j}{2}$ 31. $-3, 3$ 32. $0, 7$

33. $1, -9$ 34. $\dfrac{-3 \pm \sqrt{29}}{2}$ 35. $\dfrac{1 \pm \sqrt{11}}{2}$ 36. $2, 2$

37. $1 \pm 2\sqrt{2}$ 38. $-1, \dfrac{5}{2}$ 39. $\dfrac{1 \pm \sqrt{41}}{5}$ 40. $\dfrac{2}{3}, -2$

41. $2 \pm \sqrt{3}$ 42. $\dfrac{1 \pm \sqrt{37}}{6}$ 43. $\dfrac{1 \pm j\sqrt{7}}{4}$ 44. $-2, 3$

45. $-1 \pm j$ 46. $2 \pm 2\sqrt{3}$ 47. 27 48. $20, 21$

49. 4 in., 5 in. 50. 7.00 m 51. $-r \pm \sqrt{r^2 - k^2}$ 52. 1.18

53. 6 s 54. 49 A

55. $\dfrac{-v \pm \sqrt{v^2 + 2as}}{a}$ 56. $p = \dfrac{2.5 \pm \sqrt{6.25 - 50.4n}}{25.2}$

57. 5, 20 58. $x = L, 2L$ 59. 7 mm 60. 5.58 mm

61. 33 poles 62. $21.10

63. 300 mi/h 64. 1.5 ft by 2.0 ft

65. 1, -0.35 66. 0.62, -1.62

67. 0.98, -0.30 68. 6.12, 0.05

69. 6.1199 cm 70. 0.63500 cm

71. 5.16 ft 72. 3.116 m, 1.926 m

EXERCISES 12-1

1. $\log 100 = 2$ 2. $\log 1000 = 3$ 3. $\log 0.01 = -2$
4. $\log 0.0001 = -4$ 5. $\log 2884 = 3.4600$ 6. $\log 0.02692 = -1.57$
7. $\log 0.0003594 = -3.4444$ 8. $\log 4.6419 = 0.6667$
9. $\log_2 1024 = 10$ 10. $\log_3 81 = 4$ 11. $\log_6 216 = 3$
12. $\log_5 625 = 4$ 13. $10^1 = 10$ 14. $10^2 = 100$
15. $10^3 = 1000$ 16. $10^6 = 1,000,000$ 17. $10^{-2} = 0.01$
18. $10^{-4} = 0.0001$ 19. $10^{2.7536} = 567$ 20. $10^{-2.4634} = 0.00344$
21. $2^3 = 8$ 22. $2^6 = 64$ 23. $3^5 = 243$
24. $5^3 = 125$ 25. 4   26. 5
27. -3   28. -5   29. 6
30. 12   31. -3   32. -8
33. 4   34. 2   35. 3
36. 4   37. 0.8609   38. 0.9112
39. 1.9713   40. 3.4150   41. 4.9191
42. 2.3181   43. 4.6990   44. 8.4082
45. 8.7267 - 10   46. 7.5539 - 10   47. 8.4843 - 10
48. 5.8041 - 10   49. 6   50. 3
51. 2   52. 5   53. 625
54. 256   55. 10   56. 2
57. 2   58. -3   59. 100,000
60. 0.00001   61. 6.1271   62. 7.0253
63. 4.3010   64. 41.7782   65. 9.9868
66. -0.9031   67. -7.7959   68. -4.6576

EXERCISES 12-2

1. 46.3	2. 77.6	3. 54.7	4. 61.4
5. 12,600	6. 0.000450	7. 115,000	8. 578
9. 2.19	10. 1870	11. 1.36	12. 0.00137
13. 0.0296	14. 0.00339	15. 12.9	16. 1.96
17. 3.74	18. 0.278	19. 4.12	20. 3.51
21. 20.8	22. 5140	23. 0.243	24. 7,300,000
25. 5.18	26. 3.85	27. 2.38	28. 0.770
29. 1.65	30. 88.4	31. 20.1	32. 221
33. 29.2 in.^{2}	34. $968.75	35. 8.4	36. 70 dB

37. 85.4 ft^3 38. 26,200 lb   39. 2.13 V

40. 577 in./min   41. 26.7 bu/acre   42. 0.00845 m^3

43. 332 m/s   44. 7600 ft^3 45. 3,010,000 ft

46. 52.8%   47. 1.29   48. 0.0376

49. 4.56×10^{192} 50. 3.375×10^{1849} 51. 3.27×10^{150}

52. 3.94×10^{115} 53. 1.34×10^{154} 54. 6.51×10^{50}

55. 3.76×10^{414} 56. 4.24×10^{194}

EXERCISES 12-3

1. 2.1282	2. 2.8904	3. -0.1054	4. -1.6094
5. 5.2983	6. 6.4615	7. -0.9808	8. 6.9078
9. -6.9078	10. -4.4228	11. 8.2161	12. -7.9662
13. 9.90	14. 0.61	15. 304.2	16. 6.03
17. 28.1	18. 3.050	19. 5.46×10^{18}	20. 0.000239
21. 2.7080	22. -0.5108	23. 4.8282	24. 5.4930
25. 4.3944	26. 3.2188	27. 7.6132	28. 1.1756

29. 3.0910	30. 4.46	31. 15,000 years	32. 2150 J
33. 3.04 years	34. 550 μf	35. 674 years	36. 9.90%
37. 0.0927 s	38. 0.384 s	39. 8.50%	40. 6.2×10^9 Ω
41. 6.3363	42. -5.5662	43. 248.4547	44. 54.7549
45. 317	46. 1.79	47. 0.6709	48. 0.118073

EXERCISES 12-4

1. 0.8528	2. 3.6420	3. -0.8147	4. -2.2199
5. 1.4050	6. 2.4922	7. 4.4970	8. -2.2198
9. 3.025	10. 6.045	11. 70.34	12. 184.8
13. 3357	14. 0.7423	15. 2.862×10^{-5}	16. 0.1273
17. 1443 m/s	18. 1554 m/s	19. 1481 m/s	20. 1587 m/s
21. 4°	22. $57^{\circ}C$	23. 46°	24. $34^{\circ}C$
25. 445 mg	26. 434 mg	27. 228 mg	28. 200 mg
29. 45 days	30. 59 days	31. 131 days	32. 141 days
33. 46.77	34. 5.229	35. 3.483×10^7	36. 0.01413
37. 106.2 lb	38. 389.2	39. 1.970×10^8 mi^2	40. 2.565 A

EXERCISES 12-5

1. $\log 10,000 = 4$	2. $\log 0.001 = -3$	3. $\log_6 1296 = 4$
4. $\log_5 0.008 = -3$	5. $\log 10 = 1$	6. $\log 17.0 = 1.23$
7. $\log 0.00288 = -2.54$		8. $\log 2.5 = 0.4$
9. $2^7 = 128$	10. $3^4 = 81$	11. $10^5 = 100,000$
12. $10^{-1} = 0.1$	13. $e^{3.0} = 20$	14. $e^{-1.4} = 0.25$
15. $4^5 = 1024$	16. $5^{-2} = 0.04$	17. 7.14
18. 1910	19. 1.79	20. 641

21. 1.77×10^{-5} 22. 1.17×10^7 23. 153.9

24. 0.2398   25. 3.43   26. 77.7

27. 30.9   28. 0.0442   29. 7.67×10^4

30. 0.495   31. 9.716   32. 16,670

33. 1260   34. 2.16   35. 3.35×10^7

36. 0.0109   37. 1462   38. 1.010

39. 4.81   40. 9.21   41. 1.05

42. 9.24   43. 36.17   44. 0.9657

45. 692   46. 6.40   47. 35.2

48. 322.4   49. 1.4110   50. 2.4849

51. 9.4892 - 10   52. 7.6974-10   53. 1.7917

54. 2.0793   55. 0.4055   56. 3.4655

57. 3.7478   58. 4.2761   59. 4.5949

60. 2.2061   61. 2.59   62. 1.6×10^5

63. 140 dB   64. 1370   65. 50%

66. 3.92×10^{-4} 67. 1.0×10^{-7} 68. 15.8°C

69. 4.60   70. 4.69 kg   71. 152 mi

72. 499 s   73. $13,900   74. 873.3 ft·lb

75. 162 ft/s   76. 10.8 km/s   77. 736.0 m²

78. 6.16 atm   79. 0.0200 s   80. 2048

81. 1.5750   82. -1.4823   83. 2.7840

84. -7.0274   85. 221.651   86. 0.003456002

87. 3.83604   88. 0.36586   89. 1.26×10^{120}

90. 1.91×10^{119} 91. 1.04×10^{159} 92. 1.16×10^{-45}

EXERCISES 13-1

1. y is dependent variable, x is independent variable.
2. s is dependent variable, t is independent variable.
3. p is dependent variable, V is independent variable.
4. v is dependent variable, t is independent variable.
5. Multiply value of independent variable by 3.
6. Add 3 to the value of the independent variable.
7. Subtract the square of the independent variable from 2.
8. Add the square of the independent variable to the value of the independent variable.

9. $f(x) = 5 - x$ 10. $f(q) = 6q^2$ 11. $f(t) = t^2 - 3t$

12. $f(r) = 10^{-5r}$ 13. 0, 3   14. 2, -1

15. 7, -5   16. -1, 5   17. -2, $-\dfrac{5}{4}$

18. 2, 6   19. -1, 2.91   20. -5, -33

21. -1, 8   22. $a^3 + 2a - 1$ 23. 2, 2

24. -54, -108   25. $\dfrac{1}{2}$, undefined   26. $\dfrac{26}{5}$, undefined

27. $a^2 - 2a^4$; $\dfrac{a-2}{a^2}$ 28. $\dfrac{1}{16}$, 2   29. Function

30. Function   31. Not a function   32. Not a function

33. $c = 2\pi r$ 34. $V = e^3$ 35. $A = \dfrac{p^2}{16}$

36. $A = \dfrac{1}{4}\pi d^2$ 37. $t = 0.02n$ 38. $C = P + 0.5P$

39. $s = 3v$ 40. $C = 500 + 5(l - 50)$ 41. $R_1 = R_2 + 1500$

42. $p_1 = 2p_2 - 5000$ 43. $i = \dfrac{V}{5}$ 44. $d = \dfrac{1}{125}s^2$

EXERCISES 13-2

1. $A(2, 1)$, $B(-2, 3)$ 2. $C(5, 2)$, $D(-2, -2)$

3. $E(4, 0)$, $F(-2, 1)$ 4. $G(3, -4)$, $H(-5, -1)$

5. $I(1, 5.5)$, $J(3, 5.5)$ 6. $K(0, -7)$, $L(-4, -10)$

7. $M(-9.5, 0)$, $N(-9.5, 2)$ 8. $P(8, -7)$, $Q(-5, 9)$

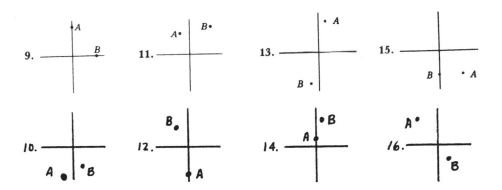

9.   11.   13.   15.

10.   12.   14.   16.

17. I, II   18. IV, III   19. III, II   20. IV, II

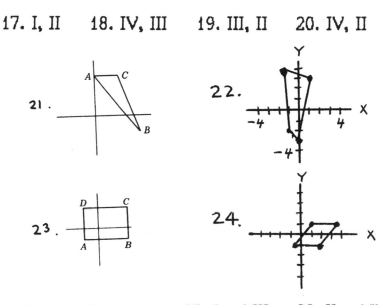

21.

22.

23.

24.

25. 0   26. 0   27. I and III   28. II and IV

29. IV   30. II

31. (1.25, 1.25), (1.25, -1.25), (-1.25, 1.25), (-1.25, -1.25)

32. Right triangle   33. Isosceles   34. (1, 2)

35. 3   36. Straight line bisecting first and third quadrants.

EXERCISES 13-3

1.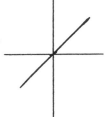

2.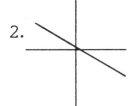

3.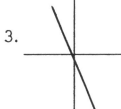

4.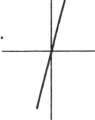

5.

6.

7.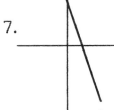

8.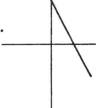

9.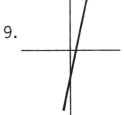

10.

11.

12.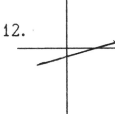

13.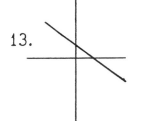

14.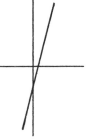

15.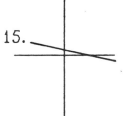

16.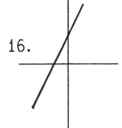

17.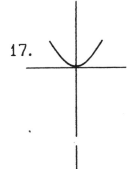

18.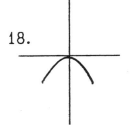

19.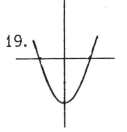

20.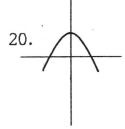

21.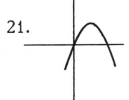

22.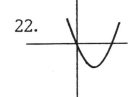

23.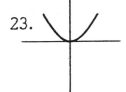

24.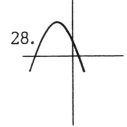

25.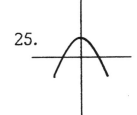

26.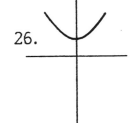

27.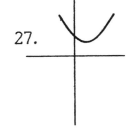

28.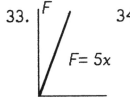

29. (0, 3), (-3, 0)

30. (0, 8), (-4, 0)

31. (0, 6), (2, 0)

32. (0, -8), (-1.6, 0)

33.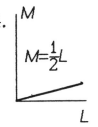

$F = 5x$

34.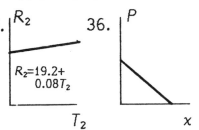

$M = \frac{1}{2}L$

35. $R_2 = 19.2 + 0.08T_2$

36. P, x

37.

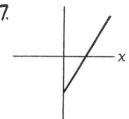

$p = 1.5x - 200$

38. F

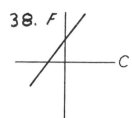

C

39.

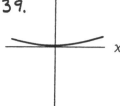

40. y

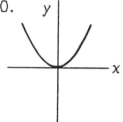

x

41. t

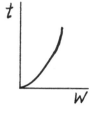

W

42. A

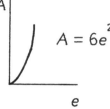

$A = 6e^2$

e

43. h

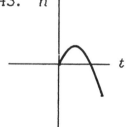

t

44. y

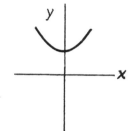

x

104

EXERCISES 13-4

1.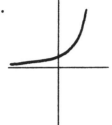

2.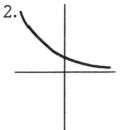

3.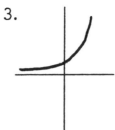

4.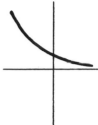

5.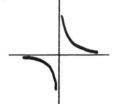

6.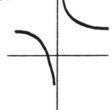

7.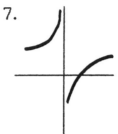

8.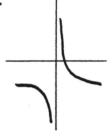

9.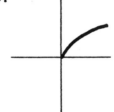

10.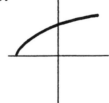

11.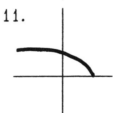

12.

13.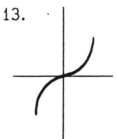

14.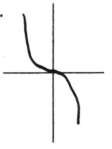

15.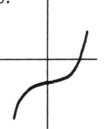

16.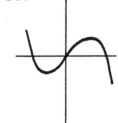

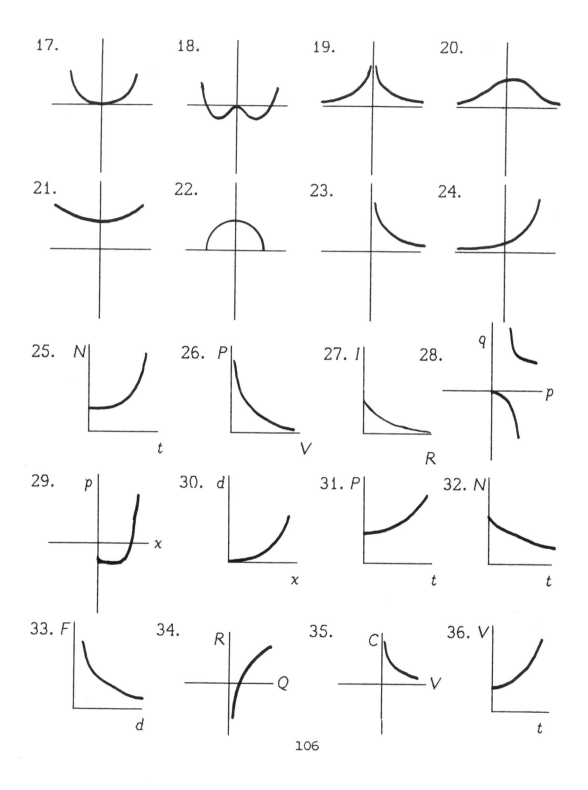

17.

18.

19.

20.

21.

22.

23.

24.

25. N — t

26. P — V

27. I — R

28. q — p

29. p — x

30. d — x

31. P — t

32. N — t

33. F — d

34. R — Q

35. C — V

36. V — t

EXERCISES 13-5

1. 1.4, -1.2   2. -1, 1.5   3. 9, -8   4. -0.14, 0.29

5. 6.5, 11.6   6. 4.5, 1.8   7. 9.2, 1.7   8. -1.2, -1.2

9. 8.9, 2.1   10. -3.9, 1.6   11. -2.1, 5.0   12. 2.4, 3.2

13. 0.7   14. -4.3   15. -1.2, 1.7   16. 1.2, -4.2

17. 1.7   18. -1.9   19. 4.2   20. 1.4

21. 3.5   22. 0.5   23. 1.5   24. 0.5

25. 0.0065 C, 0.0001 C   26. 2.6 cm, 3.8 cm

27. 35 g, 70 g   28. 40 mg, 270 mg

29. 340 m/s, 346 m/s, 354 m/s, 360 m/s

30. -4.0 in., -5.0 in., -5.0 in., -2.3 in.

31. 800 Ω, 78 Ω, 15 Ω, 11 Ω

32. 14 ft/s, 19 ft/s, 20 ft/s, 22 ft/s

33. 28 lb   34. 4.8 s   35. 2.1 in.   36. 1.26 μs

EXERCISES 13-6

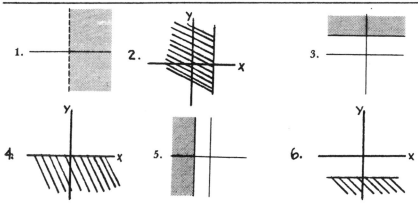

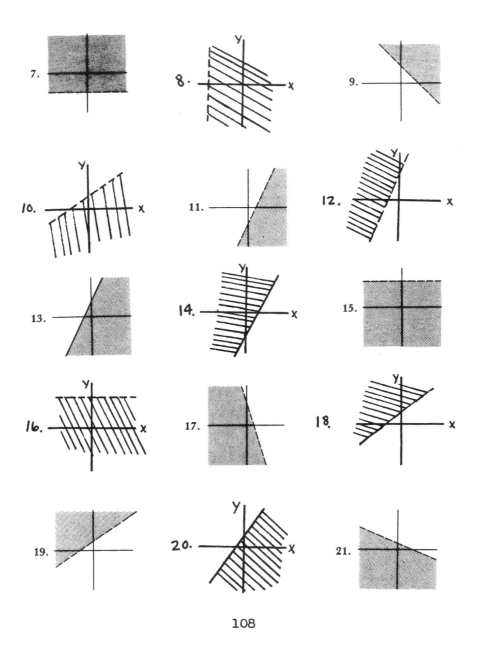

7.

8.

9.

10.

11.

12.

13.

14.

15.

16.

17.

18.

19.

20.

21.

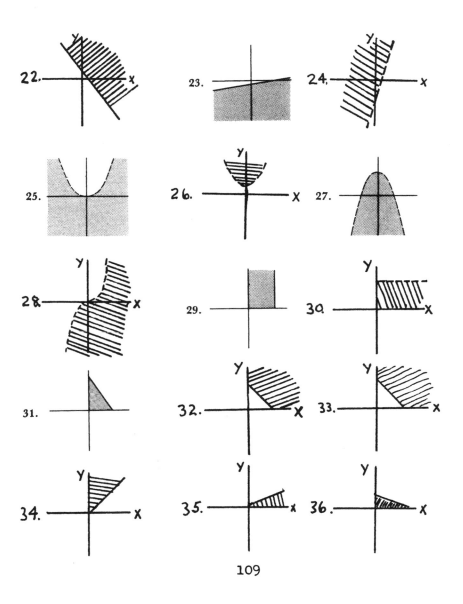

22.

23.

24.

25.

26.

27.

28.

29.

30.

31.

32.

33.

34.

35.

36.

EXERCISES 13-7

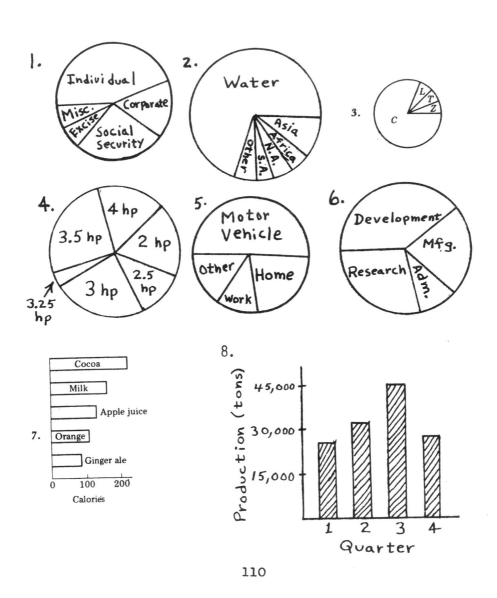

1. Individual / Corporate / Social Security / Excise / Misc.

2. Water / Asia / Africa / N.A. / S.A. / Other

3. C / L / T / Z

4. 4 hp / 3.5 hp / 2 hp / 2.5 hp / 3 hp / 3.25 hp

5. Motor Vehicle / Other / Home / Work

6. Development / Mfg. / Research / Adm.

7. Cocoa / Milk / Apple juice / Orange / Ginger ale
0   100   200
Calories

8. Production (tons)
45,000
30,000
15,000
1 2 3 4
Quarter

110

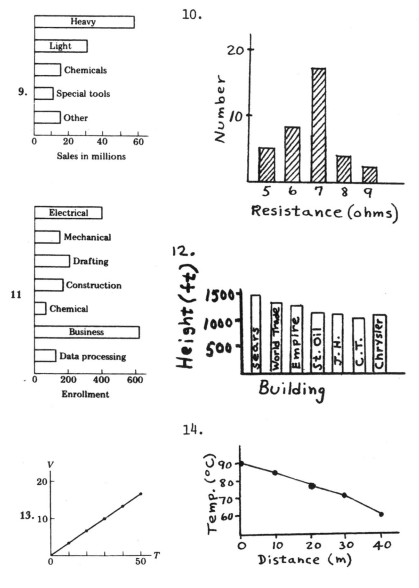

9.

Heavy

Light

Chemicals

Special tools

Other

| 0 | 20 | 40 | 60 |

Sales in millions

10.

Number

20

10

5 6 7 8 9

Resistance (ohms)

11

Electrical

Mechanical

Drafting

Construction

Chemical

Business

Data processing

| 0 | 200 | 400 | 600 |

Enrollment

12.

Height (ft)

1500

1000

500

Sears World Trade Empire St. Oil J.H. C.T. Chrysler

Building

13.

V

20

10

0   50   T

14.

Temp. (°C)

90

80

70

60

0   10   20   30   40

Distance (m)

111

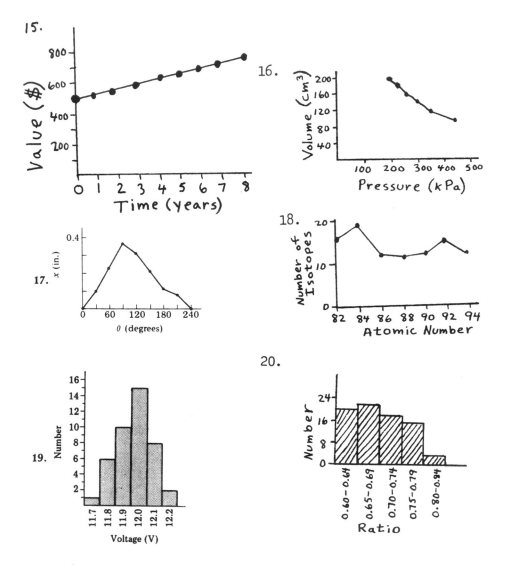

15.

Value ($) vs Time (years)

16.

Volume (cm³) vs Pressure (kPa)

17.

x (in.) vs θ (degrees)

18.

Number of Isotopes vs Atomic Number

19.

Number vs Voltage (V)

20.

Number vs Ratio

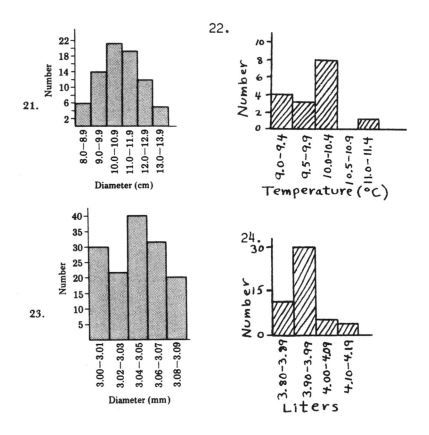

21.

22.

23.

24.

EXERCISES 13-8

1. 3, 2

2. -3, 3

3. 4, $\frac{5}{3}$

4. -22, -10

5. 3, -$\frac{1}{2}$

6. -9, -16

7. 12, -3.36

8. 4, 7 + v -4v²

9. 0, 8

10. -16, 9

11. 1, 5

12. $\frac{2}{3}$, 2

13.

14.

15.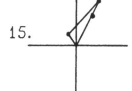

16.

17.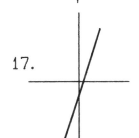

18.

19.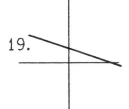

20.

21.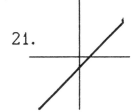

22.

23.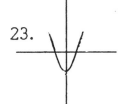

24.

25.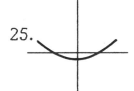

26.

27.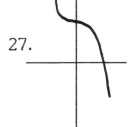

28.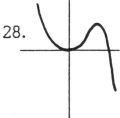

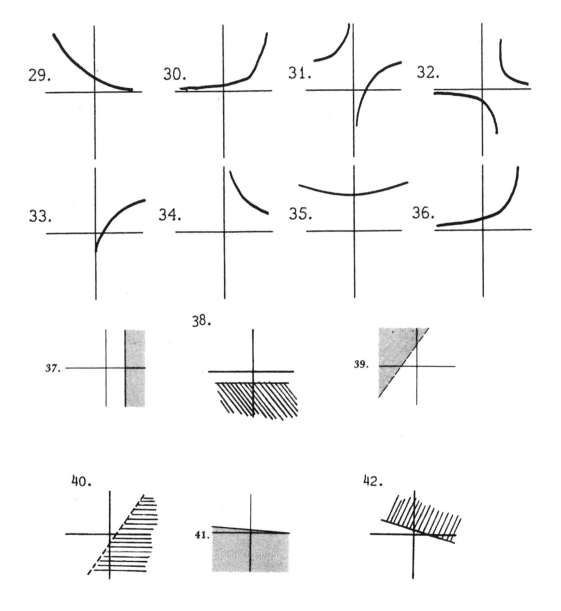

29.

30.

31.

32.

33.

34.

35.

36.

38.

37.

39.

40.

41.

42.

115

43.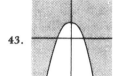

44.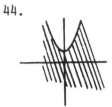

45.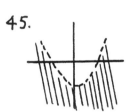

46.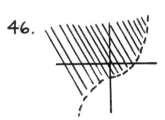

47. 1.3

48. -1.8

49. -0.7, 1.0

50. 0.7, -7.7

51. 0.6

52. 1.0, 1.2

53.

54.

55.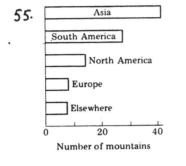

56.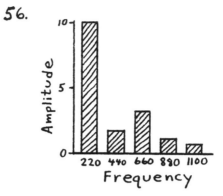

57.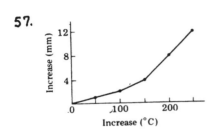

58.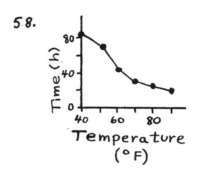

59.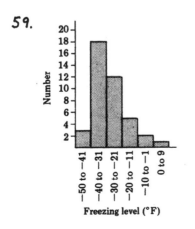

60.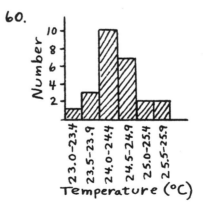

61. $A = 3x$ 62. $p = 2x + 8$ 63. $T = 0.06C$

64. $I = 200 + 0.03\ S$ 65. $H = 240I^2$

66. $V = 2A$ 67. $A = \frac{1}{2}\pi r^2 + 4r$ 68. $T = \dfrac{160}{\frac{I}{2}}$

69. $700, $1700, $2400   70. 4.0 mi/s, 4.2 mi/s, 4.9 mi/s

71. 101 ft, 103 ft, 112 ft   72. 67 m

73.
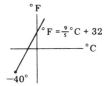

74.

75.
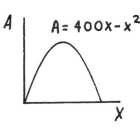

76.
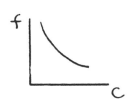

118

EXERCISES 14-1

1. $x = 3.0$, $y = 0.0$ 2. $x = 2.0$, $y = 3.0$

3. $x = 2.0$, $y = 1.0$ 4. $x = 1.0$, $y = 2.0$

5. $r = 4.0$, $x = -3.0$ 6. $m = -2.0$, $n = 8.0$

7. $x = 2.0$, $y = 2.0$ 8. $x = 0.9$, $y = 0.1$

9. $x = -1.0$, $y = 2.0$ 10. $R = 3.0$, $T = -1.0$

11. $x = -2.0$, $y = 1.5$ 12. $x = -0.3$, $y = -2.5$

13. $a = -1.9$, $b = -0.5$ 14. $u = 4.5$, $v = 0.4$

15. Inconsistent   16. Dependent

17. $p = 2.4$, $q = 1.1$ 18. $x = 1.6$, $y = -0.9$

19. $x = 1.9$, $y = -2.2$ 20. $r = 2.4$, $s = 1.4$

21. $x = 5.9$, $y = -0.2$ 22. $x = 4.1$, $y = -0.1$

23. $x = 0.8$, $y = 3.6$ 24. $x = 0.4$, $y = -3.9$

25. Inconsistent   26. Dependent

27. Dependent   28. Inconsistent

29. Dependent   30. Inconsistent

31. Inconsistent   32. Dependent

33. $x = 95$, $y = 25$ 34. $P = 110$, $W = 30$

35. $x = 33$ lb, $y = 67$ lb   36. 1.8 min

37. $r_1 = 3.5$ mi/h, $r_2 = 1.5$ mi/h

38. $g = \$6.00$ per hour, $d = \$4.00$ per hour

39. $R_1 = 2.5\ \Omega$, $R_2 = 1.0\ \Omega$ 40. $T_1 = 12.2$ lb, $T_2 = 14.3$ lb

EXERCISES 14-2

1. $x = 5$, $y = 8$ 2. $x = 1$, $y = 3$

3. $x = 6$, $y = 4$ 4. $x = 2$, $y = -4$

5. $x = \dfrac{2}{5}, y = \dfrac{3}{5}$ 6. $x = \dfrac{1}{10}, y = \dfrac{9}{10}$

7. $x = -1, y = 2$ 8. $x = 0, y = 0$

9. $x = 2, y = 1$ 10. $a = \dfrac{16}{5}, b = -\dfrac{2}{5}$

11. $x = 0, y = 0$ 12. $x = \dfrac{18}{5}, y = -\dfrac{2}{5}$

13. $x = 6, y = 0$ 14. Inconsistent

15. $u = -\dfrac{6}{17}, k = \dfrac{20}{17}$ 16. $u = \dfrac{8}{7}, z = \dfrac{5}{7}$

17. $x = \dfrac{81}{10}, y = \dfrac{63}{10}$ 18. $x = 25, y = -\dfrac{9}{2}$

19. $x = -2, y = -1$ 20. $u = \dfrac{3}{2}, w = 1$

21. $r = 4, s = -2$ 22. $m = -1, n = 5$

23. $x = \dfrac{32}{31}, k = -\dfrac{69}{31}$ 24. $y = \dfrac{70}{31}, z = \dfrac{12}{31}$

25. $x = 1000, y = 200$ 26. $x = 300, y = 180$

27. $r_1 = 1.25, r_2 = 0.75$ 28. 60 mL

29. $V_1 = 40$ V, $V_2 = 20$ V   30. $x = \$3000, y = \$2000$

31. $b = 24$ ft, $r = 13$ ft   32. $T_1 = 630$ N, $T_2 = 560$ N

33. $x = -0.995, y = 1.36$ 34. $x = 0.58505, y = -0.49530$

35. $x = -0.2268, y = 1.873$ 36. $x = 0.550, y = 0.143$

EXERCISES 14-3

1. $x = 5, y = 2$ 2. $x = \dfrac{11}{3}, y = \dfrac{4}{3}$

3. $x = 2, y = 1$ 4. $x = 4, y = 7$

5. $m = 4, n = 8$ 6. $x = 2, y = 5$

7. $d = -1, t = 4$ 8. $r = 1, s = -1$

9. $x = 1, n = 5$
10. $x = 2, y = -2$
11. $x = 9, y = 2$
12. $t = 2, x = 8$
13. Inconsistent
14. Dependent
15. $a = 1, b = 2$
16. $x = -\frac{73}{19}, y = -\frac{78}{19}$
17. $p = \frac{3}{2}, q = -\frac{1}{12}$
18. $k = -4, t = 3$
19. $x = -14, y = -3$
20. $x = -16, y = 24$
21. $m = -4, n = 6$
22. $x = \frac{146}{201}, y = \frac{409}{804}$
23. $x = 1, y = 1$
24. $x = 2, y = 1$
25. $V_1 = 6.0$ V, $V_2 = 1.5$ V
26. $x = 8$ ft, $y = 5$ ft
27. $x = \frac{1}{16}, y = 1$
28. $x = 2000$ lines/min, $y = 3500$ lines/min
29. $c_1 = 10$ ft, $c_2 = 4$ ft
30. $t_1 = 5$ h, $t_2 = 4$ h
31. 650 math, 500 verbal
32. $x = \$2500, y = \$4000$
33. $m = 15$ L, $c = 5$ L
34. $I_1 = -\frac{1}{3}$ A, $I_2 = 1$ A
35. $x = 18$ lb, $y = 10$ lb
36. $x = 11, y = 6$
37. $x = -17.0, y = -22.5$
38. $x = 9.43, y = -5.43$
39. $x = 3.666, y = -0.4703$
40. $x = -0.2452, y = 0.3539$

EXERCISES 14-4

1. -2	2. 10	3. 31	4. -18
5. 0	6. 14	7. -2	8. -50
9. 6	10. 24	11. 52	12. -178

13. $x = 1, y = 5$
14. $x = 1, y = 3$
15. $x = 2, y = 3$
16. $x = 1, y = -2$

17. $s = 4$, $t = 3$　　　　　　　18. $R_1 = 3$, $R_2 = 8$

19. $v_1 = 5$, $v_2 = 3$　　　　　20. $m = -4$, $n = 4$

21. $x = -6$, $y = 7$　　　　　　22. $x = -4$, $y = 9$

23. $x = 12$, $y = 9$　　　　　　24. $x = 8$, $y = 6$

25. $x = -12$, $y = -10$　　　　26. $x = -15$, $y = 8$

27. $x = 3$, $y = \frac{1}{2}$　　　　　　28. $x = \frac{1}{5}$, $y = 2$

29. $x = 0.2$, $y = 0.3$　　　　　30. $x = 0.5$, $y = -0.3$

31. $x = \frac{1}{2}$, $y = \frac{1}{4}$　　　　　32. $x = \frac{1}{3}$, $y = \frac{1}{2}$

33. $R_1 = 24{,}000\ \Omega$, $R_2 = 8000\ \Omega$　34. $x = 200$, $y = 210$

35. $x = 5.3$, $y = 10.7$　　　　36. $x = \$6000$, $y = \$2000$

37. $f = \$4.50$, $b = \$0.75$　　38. $p = \$3.50$, $f = \$5.00$

39. $R_1 = 40.2\ \Omega$, $R_2 = 15.5\ \Omega$　40. $b = 26.0$ ft, $r = 14.6$ ft

41. $x = 1.43$, $y = -0.653$　　42. $x = -1.47$, $y = 0.397$

43. $x = 0.006088$, $y = 0.001498$　44. $x = 173.8$, $y = -90.27$

EXERCISES 14-5

1. $x = 1$, $y = 2$, $z = 3$　　　　2. $x = 4$, $y = 0$, $z = 1$

3. $x = -1$, $y = 2$, $z = 2$　　　4. $x = 3$, $y = -1$, $x = 5$

5. $x = 1$, $y = -1$, $z = 2$　　　6. $x = 3$, $y = 0$, $z = 1$

7. $x = 5$, $y = 4$, $z = -1$　　　8. $x = -1$, $y = -2$, $z = 2$

9. $x = 4$, $y = -1$, $z = 3$　　　10. $x = 2$, $y = 2$, $z = 2$

11. $x = 1$, $y = 1$, $z = 1$　　　12. $x = 0$, $y = 0$, $z = 0$

13. $x = 8$, $y = 5$, $z = -2$　　14. $x = 10$, $y = -5$, $z = 1$

15. $x = 7$, $y = -1$, $z = -2$　　16. $x = 8$, $y = 3$, $z = 3$

17. $x = 10$, $y = 12$, $z = -8$　18. $x = 2$, $y = 4$, $z = 6$

19. $x = 1$, $y = 2$, $z = -8$ 20. $x = 7$, $y = 5$, $z = 9$
21. $x = 4$, $y = 8$, $z = 12$ 22. $x = 6$, $y = 9$, $z = -2$
23. $x = 4$, $y = 6$, $z = -3$ 24. $x = 6$, $y = -6$, $z = 12$
25. $I_1 = 1$ A, $I_2 = 5$ A, $I_3 = 6$ A
26. $F_x = 25$ lb, $F_y = 16$ lb, $F_z = 20$ lb

27. $x = 10$ cm^3, $y = 50$ cm^3, $z = 20$ cm^3

28. $r_1 = 120$ gal/min, $r_2 = 80$ gal/min, $r_3 = 200$ gal/min

29. $r_1 = 12$ mi/gal, $r_2 = 16$ mi/gal, $r_3 = 24$ mi/gal

30. $x = 100$, $y = 130$, $z = 180$

31. $x = 0.05$, $y = 0.08$, $z = 0.06$ 32. $B = \$200$, $P = \$35$, $G = \$15$

EXERCISES 14-6

1. 8 A, -3 A   2. $A = 143$ ft^2, $B = 109$ ft^2
3. 10.5 ft from one end   4. 70 V, 30 V

5. 75 h, 30 h   6. $\frac{100}{3}$ kW, $\frac{50}{3}$ kW

7. 8 days, 14 days   8. 8 mm, 21 mm
9. $35,000 at 8%; $10,000 at 6%
10. $8000 at 4%; $200 at 5%   11. 400 ft by 500 ft
12. 2910 gal, 1920 gal   13. 90,000; 60,000
14. $380 copier, $120 computer
15. 1 year; 1.5 years   16. 0.1 h, 0.7 h
17. 42%   18. -40°C
19. 5 Ω, 9 Ω   20. 40 mi/h, 45 mi/h
21. 8 h, 10 h   22. 10 dimes, 10 quarters

23. 16, 18

24. $2.50, $1.50

25. 80 g, 40 g

26. 9.6 L

27. 15.8 lb

28. No

29. 30 A, 10 A, 80 A

30. 2000 in research, 300 in sales, 1200 in manufacturing

31. 20 gal/min, 60 gal/min, 10 gal/min

32. 5 cm³ copper, 10 cm³ nickel, 40 cm³ zinc

EXERCISES 14-7

1. $x = \frac{12}{5}, y = \frac{6}{5}$

2. $x = 6, y = -6$

3. $x = 4, y = 4$

4. $x = 4, y = 4$

5. $x = 4, y = 4$

6. $x = -\frac{6}{11}, y = -\frac{21}{11}$

7. $p = \frac{18}{5}, q = \frac{2}{5}$

8. $m = \frac{7}{2}, n = \frac{1}{2}$

9. $u = 1, v = -1$

10. $x = 2, y = -3$

11. $a = 12, b = -3$

12. $p = -\frac{7}{2}, x = -2$

13. $x = -9, y = 11$

14. $h = \frac{6}{5}, y = \frac{5}{3}$

15. $y = 1, z = -\frac{1}{2}$

16. $x = -\frac{3}{17}, y = \frac{8}{17}$

17. Dependent

18. Inconsistent

19. $x = -\frac{7}{3}, y = 2$

20. $x = -\frac{1}{4}, y = -8$

21. $x = \frac{43}{19}, y = -\frac{22}{19}$

22. $x = \frac{30}{59}, y = \frac{7}{59}$

23. $s = \frac{94}{107}, t = -\frac{22}{107}$

24. $x = -\frac{47}{191}, y = -\frac{301}{191}$

25. $x = 100$, $y = -1$

26. $x = \dfrac{1400}{43}$, $y = \dfrac{220}{43}$

27. $x = \dfrac{119}{201}$, $y = -\dfrac{59}{201}$

28. $r = \dfrac{3}{2}$, $t = \dfrac{1}{3}$

29. $x = 12$, $y = 24$

30. $x = \dfrac{180}{23}$, $y = \dfrac{132}{23}$

31. $r = \dfrac{1}{3}$, $s = \dfrac{1}{5}$

32. $x = \dfrac{1}{4}$, $y = \dfrac{1}{2}$

33. $x = 1.0$, $y = 0.3$

34. $x = 6.0$, $y = -4.0$

35. $x = 1.6$, $y = -1.2$

36. $x = 1.7$, $y = -2.3$

37. $u = 2.0$, $v = -6.0$

38. $r = 4.0$, $s = 0.5$

39. Inconsistent

40. $m = 2.6$, $n = 1.5$

41. -2

42. -28

43. -7

44. -159

45. $x = -7$, $y = 2$

46. $x = 1$, $y = -8$

47. $x = \dfrac{1}{2}$, $y = 4$

48. $x = 4$, $y = -6$

49. $x = 2$, $y = -2$, $z = 3$

50. $x = 1$, $x = 5$, $z = -2$

51. $x = 6$, $y = 4$, $z = -3$

52. $x = 5$, $y = -6$, $z = 8$

53. $i_1 = -\dfrac{3}{22}$ A, $i_2 = -\dfrac{39}{110}$ A

54. $v_0 = 6$ ft/s, $a = 10$ ft/s^2

55. $m = 416$ mol/h, $n = 331$ mol/h

56. $T_2 = 25.0$ lb, $T_3 = 43.3$ lb

57. $I_1 = 5$ A, $I_2 = 2$ A, $I_3 = -7$ A

58. $t_1 = 36$ h, $t_2 = 4$ h, $t_3 = 8$ h

59. $x = 600$ g, $y = 400$ g, $z = 80$ g

60. $x = \$2500$, $y = \$4500$, $z = \$1000$

61. 45 spots/h, 30 spots/h

62. 48, 72

63. $d = 40.0$ ft, $s = 30.0$ ft

64. 9 km, 3 km

65. 72 refunds, 112 additional

66. 97°C, 49°C

67. \$300 fixed cost, 25¢ per booklet

68. 8 ft, 10 ft

69. 45 mi/h, 55 mi/h   70. 175 km/h, 25 km/h

71. 750 mL of 5% solution; 250 mL of 25% solution

72. 31.7 mL, 63.3 mL

73. 15 Ω, 20 Ω, 40 Ω 74. 20, 12, 8

75. 12 nuts, 12 bolts, 24 washers

76. 1000, 2000, 6000

77. $x = -30.3, -28.5$ 78. $x = 0.2190, y = -0.7602$

79. $x = 0.6348, 0.03160$ 80. $x = 0.07065, y = 0.7586$

EXERCISES 15-1

1. 53° 2. 7° 3. 21° 4. 153°

5. $\angle BEC$ or $\angle CED$ 6. $\angle AED$

7. $\angle AEB$ and $\angle BEC$; $\angle BEC$ and $\angle CED$; $\angle AEC$ and $\angle CED$; $\angle AEB$ and $\angle BED$ (any two pairs)   8. $\angle AEB$ and $\angle BEC$

9. $\angle CBE$ and $\angle EBA$

10. $\angle CBD$ and $\angle DBE$; $\angle CBD$ and $\angle DBA$; $\angle DBE$ and $\angle EBA$; $\angle CBE$ and $\angle EBA$ (any two pairs)

11. $\angle CBD$ 12. $\angle DBA$ 13. 25° 14. 65°

15. 115° 16. 25° 17. 40° 18. 140°

19. 62° 20. 118° 21. $\angle 1$ and $\angle 5$; $\angle 3$ and $\angle 4$

22. $\angle 3$ and $\angle 5$ 23. $\angle 1$ and $\angle 3$ or $\angle 4$ and $\angle 5$

24. $\angle 1$ and $\angle 2$; $\angle 3$ and $\angle 2$; $\angle 4$ and $\angle 2$; $\angle 5$ and $\angle 2$ (any pair)

25. 50° 26. 130° 27. 130° 28. 50°

29. 58° 30. 32° 31. 148° 32. 58°

33. 40° 34. 50° 35. 50° 36. 40°

37. 40° 38. 80° 39. 100° 40. 80°

41. $37^o22'18''$ 42. $89^o1'4''$ 43. $7^o13''$ 44. $175^o47'57''$

126

EXERCISES 15-2

1. 56°	2. 42°	3. 48°	4. 35°
5. 68°	6. 105°	7. 80°	8. 102°
9. 60°	10. 45°	11. 120°	12. 70°
13. 32°	14. 118°	15. 76°	16. 90°
17. *AE, GD*	18. *AF*	19. *GC, BC*	20. ∠*BCG*
21. ∠*BOG*	22. ∠*COG*	23. $\overparen{BG}$, $\overparen{GC}$	24. $\overparen{GBC}$, $\overparen{BCG}$
25. 60°	26. 120°	27. 30°	28. 30°
29. 110°	30. 290°	31. 35°	32. 55°
33. 3	34. 3	35. 5	36. 5
37. 72°, 72°	38. 40°	39. 60°, 60°, 60°	40. 92.4°
41. 135° each	42. 108°	43. 138°	44. 70°
45. 117°	46. 70°	47. Horizontally	48. 5°

EXERCISES 15-3

1. 5	2. 15	3. 17	4. 26
5. 8	6. 3.46	7. 4.90	8. 8.49
9. 10.6	10. 15.0	11. 28.3	12. 35.7
13. 59.9	14. 193	15. 40.9	16. 23.9
17. 2.67	18. 0.117	19. 39.1	20. 44.5
21. 5.66 cm	22. 9.66 in.	23. p = 37.9 in., A = 55.4 in.²	
24. 12.0 m	25. 10,400 ft	26. 7.9 m	27. 12.5 ft
28. 3060 ft	29. 206	30. 27.0 m	31. 521 ft
32. 12.6 mi	33. 4.73 m	34. 603 m	35. 17.3 ft
36. 36.8 ft	37. 47.0 cm	38. 18.9 Ω	39. 30.4 kV/m
40. 894 lb	41. 30.7 cm	42. 8.42 m	43. 990 ft

44. 162 ft   45. 29.7 km   46. 56.1 cm   47. 28.3 ft

48. $\frac{s\sqrt{2}}{2}$ 49. $h = \frac{s\sqrt{3}}{2}$ 50. $d = \sqrt{2}$ s

51. 562 mi   52. 1.7 m

EXERCISES 15-4

1. $20°$, $100°$, $60°$ 2. 18   3. 16.5, 19.5

4. $70°$ 5. $40°$, $65°$, $75°$ 6. 6.7, 8.3

7. 12.7, $70°$, $70°$ 8. 5.88, 8.26   9. Similar

10. Congruent   11. Neither   12. Similar

13. $\angle E$, side EF 14. $\angle C$, side BC 15. $\angle U$, side ST

16. $\angle R$, side PQ 17. 10   18. 14

19. 12   20. 15   21. 21

22. 9   23. 20   24. 16

25. 16   26. 4   27. 8

28. 16   29. 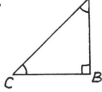   30.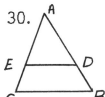

31. 3.3 cm, 4.0 cm

32. 11.7 in., 7.62 in.,
    $50°$, $100°$, $30°$

33. $\angle XKY = \angle NKF$, $\angle KXY = \angle KNF$, $\angle XYK = \angle NFK$

34. $\angle ADC = \angle ACB = 90°$; $\angle CAB = \angle CAD$. Since the two angles are
    respectively equal, all three angles must be respectively equal.
    Therefore the triangles are similar.

35. 8.0 in.   36. $140°$ 37. 6.7 ft

38. 27 m   39. 350 cm   40. 17.5 in.

41. $2\frac{1}{4}$ in.   42. 63 mi   43. 1200 km

44. 2.5 cm   45. 1530 km, 1800 km, 760 km

46. 3.7 mi   47. 20.7 m   48. 98.4 in., 98.4 in., 42.0 in.

49. 5.3 ft   50. 480 m   51. 104 m   52. 48 mi

53. 23.3 ft   54. 6.6 ft²   55. 17 cm   56. $\frac{25}{9}$

EXERCISES 15-5

1. 6900 ft³   2. 199 m³   3. 99,000 cm³

4. 347,000 in.³   5. 8.85 ft³   6. 13,500 mm³

7. 22,300 cm³   8. 1.97 yd³

9. (a) 795 ft³; (b) 1170 ft³   10. (a) 2150 mm²; (b) 6800 mm²

11. (a) 2.69 m²; (b) 4.03 m²   12. (a) 29.8 ft²; (b) 44.7 ft²

13. (a) 12,300 in.²; (b) 18,900 in.²

14. (a) 18,100 cm²; (b) 32,800 cm²

15. (a) 57.4 cm²; (b) 85.0 cm²   16. (a) 154 in.²; (b) 333 in.²

17. 18,800 cm³   18. 615 in.³   19. 2540 ft³

20. 66.5 m³   21. 14,700,000 mm³ 22. 0.0492 yd³

23. 14,800 in.³   24. 4640 cm³

25. (a) 226,000 mm²; (b) 791,000 mm²

26. (a) 30,100 ft²; (b) 52,800 ft²   27. (a) 124 in.²; (b) 546 in.²

28. (a) 15.9 m²; (b) 49.1 m²   29. (a) 641 cm²; (b) 1550 cm²

30. (a) 89.5 in.²; (b) 112 in.²   31. (a) 3360 ft²; (b) 15,100 ft²

32. (a) 32,500 mm²; (b) 43,700 mm²

33. 5.50 m³   34. 22,000 ft³   35. 594 in.²

36. 114 cm²   37. 56.6 cm³   38. 175,000 g

39. 123,000 bu   40. $1200   41. 18,100 cm²

42. 271 L   43. 600 ft³   44. 558,000 gal

EXERCISES 15-6

1. 54,000 ft³   2. 100,000 cm³   3. 958 mm³
4. 117 ft³   5. 76.5 yd²   6. 154,000 cm²
7. 240 m²   8. 2340 in.²   9. 1470 ft³
10. 26,200 cm³   11. 27,100 cm³   12. 1850 ft³
13. 198 in.²   14. 7.91 m²   15. 7210 cm²
16. 386 ft²   17. 113 ft³   18. 33,500 cm³
19. 5, 570, 000 mm³   20. 24,800 ft³
21. 11,300 in.²   22. 45,200 m²   23. 376,000 mm²
24. 12,200 in.²   25. 3,330,000 yd³   26. 263 g
27. $191   28. 298 ft²   29. 58.6 mm³
30. 1200 ft³   31. 23,500 lb   32. 94.2 m³
33. 22.3 in.²   34. 155 cm²   35. 78.8 lb
36. 447 in.³   37. 4.70 m³   38. 203 in.³
39. 38,000,000 km²   40. 2030 cm²
41. 1.10 in.³   42. 49,000 gal   43. 1.111 cm

44. 8.00 in.   45. $V = \dfrac{\pi d^2}{6}$ 46. $A = \pi d^2$

47. $A = 3\pi r^2$ 48. $A = \dfrac{1}{3}\pi r^2 (2r + h)$

EXERCISES 15-7

1. 61°   2. 151°   3. 90°
4. 60°   5. 27°   6. $\angle C = \angle D = 41°$
7. 21.5 cm   8. 0.870, 0.870   9. $\angle CBE$
10. $\angle ABE, \angle EBD$ 11. $\angle EBD$ 12. $\angle EBC$
13. $\angle 2$ and $\angle 5$ 14. $\angle 1$ and $\angle 3$ 15. 180°

16. 180°   17. 65°   18. 155°
19. 32°   20. 148°   21. 52°
22. 128°   23. 132°   24. 132°
25. 36°   26. 62°   27. 50°
28. 80°   29. 120°   30. 160°
31. 40°   32. 80°   33. 25°
34. 65°   35. 65°   36. 90°
37. 41   38. 50   39. 42
40. 33   41. 7.36   42. 281
43. 21.1   44. 0.414   45. 7.5
46. 2.4   47. 6.3 in.   48. 19.0 in.
49. 600 ft³   50. 40,000 cm³   51. 6190 cm³
52. 100 ft³   53. 97,700 in.³ 54. 656,000 mm³
55. 160,000 cm³   56. 31.1 in.³   57. 3990 ft³
58. 33,600 mm³   59. 24,400 cm³   60. 73,600 ft³
61. 900 ft²   62. 2320 cm²   63. 14.4 m²
64. 73.5 yd²   65. 5430 in.²   66. 44,500 mm²
67. 16,800 cm²   68. 50.6 in.²   69. 1230 ft²
70. 5260 mm²   71. 4070 cm²   72. 8490 ft²
73. 21.1 ft   74. 510 m   75. 231 cm
76. 19.0 ft   77. 32.8 cm   78. 16.1 Ω
79. Yes   80. $l = e\sqrt{3}$ 81. 56.1 m
82. 9.9 ft, 14.9 ft, 17.9 ft   83. 22.5°, 67.5°
84. 30°, 60°, 90° 85. 5.0 ft   86. 20 ft
87. 864,000 mi   88. 5.03 cm   89. 1600 cm
90. 1.50 m   91. 10.0 in.   92. 205 ft
93. 180°   94. 40°   95. 12,400 cm²

96. 37.8 ft²   97. 667,000 gal   98. 11.0 m

99. 2,550,000 cm³   100. 2480 in.³

101. 40.0 in.²   102. 4.89 m²   103. 7.07 cm³

104. 1.42×10¹³ mi³   105. 7.44 ft²

106. 961 cm²   107. 32,300 kg   108. 1010 ft³

109. 238 in.²   110. 43.1 m³   111. 0.204 ft

112. 5.95 ft³   113. 140 cm²   114. 233 gal

115. $A = 3\pi r^2 + 2\pi rh$ 116. $V = \frac{1}{3}e^2h + e^3$

117. 113 in.³   118. 5160 in.³   119. 16.9 cm

120. 6.37 cm²

EXERCISES 16-1

1. $\frac{9}{41}$, $\frac{9}{40}$, $\frac{9}{41}$
2. $\frac{40}{41}$, $\frac{40}{41}$, $\frac{40}{9}$
3. $\frac{9}{40}$, $\frac{41}{40}$, $\frac{40}{9}$

4. $\frac{41}{9}$, $\frac{41}{9}$, $\frac{41}{40}$
5. $\frac{1}{2}$, 2, $\sqrt{3}$
6. 2, $\frac{\sqrt{3}}{2}$, $\frac{\sqrt{3}}{3}$

7. $\frac{\sqrt{3}}{3}$, $\frac{1}{2}$, $\frac{2\sqrt{3}}{3}$
8. $\sqrt{3}$, $\frac{2\sqrt{3}}{3}$, $\frac{\sqrt{3}}{2}$
9. 0.624, 1.25, 1.60

10. 0.624, 1.60, 1.25   11. 0.782, 0.782, 0.798

12. 1.28, 0.798, 1.28   13. $\frac{4}{5}$, $\frac{3}{4}$

14. $\frac{15}{17}$, $\frac{17}{15}$
15. $\frac{7}{24}$, $\frac{24}{25}$
16. $\frac{9}{25}$, $\frac{25}{9}$

17. $\frac{\sqrt{2}}{2}$, 1
18. 1.5, 0.88
19. 1.4, 0.73

20. 0.87, 0.56   21. 0.63, 1.3   22. 1.71, 1.16

23. 0.679, 1.47   24. 1.162, 0.5093   25. $\frac{\sqrt{2}}{2}$

26. $\frac{\sqrt{3}}{2}$ 27. 1.4   28. 0.78

29. 0.491   30. 0.410   31. 1.04

32. 0.944   33. 0.5000   34. 1.02

35. 1.41   36. 0.408   37. cot B, sec B

38. Similar triangles; trigonometric ratios of corresponding sides
    are equal.

39. $\sin A = \frac{a}{c}$; $\cos A = \frac{b}{c}$; $\tan A = \frac{a}{b}$; $\sec A = \frac{c}{b}$; $\csc A = \frac{c}{a}$; $\cot A = \frac{b}{a}$

40. Yes. The denominators are the same, but the numerator for sin A
    is smaller.

41. 1   42. 1   43. $\frac{5}{13}, \frac{12}{13}, \frac{5}{12}$; $\frac{5}{13} \div \frac{12}{13} = \frac{5}{12}$

44. $\frac{8}{17}, \frac{15}{17}, \frac{8}{15}$

EXERCISES 16-2

1. 0.8480	2. 0.3584	3. 0.4557	4. 3.630
5. 1.500	6. 1.205	7. 0.9178	8. 0.6032
9. 0.2382	10. 2.023	11. 0.7046	12. 0.9871
13. 4.222	14. 1.331	15. 2.475	16. 0.9523
17. 32.0°	18. 12.0°	19. 18.9°	20. 37.4°
21. 60.5°	22. 51.2°	23. 61.9°	24. 81.4°
25. 52.7°	26. 53.1°	27. 18.0°	28. 22.5°
29. 84.3°	30. 71.2°	31. 58.2°	32. 62.2°
33. 1.206	34. 0.6746	35. 1.179	36. 0.9877
37. 53.1°	38. 25.4°	39. 40.9°	40. 58.0°
41. 34.6°	42. 65.4°	43. 79.4°	44. 5.0°
45. 0.72	46. 0.94	47. 1.9	48. 1.3

49. 13 V   50. 2930 ft   51. 126.5 m   52. 65.3°
53. 1.56   54. 424 ft·lb   55. 23.9 m   56. 18.8°
57. Error display   58.   3 $\boxed{1/x}$ $\boxed{\text{INV}}$ $\boxed{\text{SIN}}$

59.   2.05 $\boxed{1/x}$ $\boxed{\text{INV}}$ $\boxed{\text{COS}}$ $\boxed{\text{SIN}}$

60.   3 $\boxed{1/x}$ $\boxed{\text{INV}}$ $\boxed{\text{SIN}}$ $\boxed{\text{COS}}$ $\boxed{1/x}$

EXERCISES 16-3

1. $B = 60.0^\circ$, $b = 20.8$, $c = 24.0$ 2. $B = 45.0^\circ$, $a = 16.0$, $c = 22.6$
3. $A = 33.7^\circ$, $a = 12.5$, $b = 18.7$ 4. $A = 72.9^\circ$, $b = 4.83$, $c = 16.4$
5. $B = 13.2^\circ$, $a = 30.6$, $b = 7.17$ 6. $A = 54.3^\circ$, $a = 2.02$, $c = 2.48$
7. $A = 28.8^\circ$, $B = 61.2^\circ$, $b = 1.18$ 8. $A = 32.4^\circ$, $B = 57.6^\circ$, $c = 8.77$
9. $A = 82.6^\circ$, $B = 7.4^\circ$, $a = 44.6$ 10. $A = 80.0^\circ$, $B = 10.0^\circ$, $c = 745$
11. $A = 64.2^\circ$, $B = 25.8^\circ$, $b = 4.71$
12. $A = 52.6^\circ$, $B = 37.4^\circ$, $a = 0.255$
13. $B = 83.0^\circ$, $a = 1.88$, $c = 15.4$
14. $A = 5.5^\circ$, $a = 166$, $b = 1720$ 15. $A = 33.5^\circ$, $B = 56.5^\circ$, $c = 118$
16. $A = 74.7^\circ$, $B = 15.3^\circ$, $a = 6.89$

17. 29.1 ft   18. 29.6 m   19. 304 cm
20. 28.6 ft   21. 3.4° 22. 11%, 17 ft
23. 318 cm   24. 1410 ft   25. 85.2°
26. 18.9° 27. 4000 ft   28. 21.9 mi
29. 2.94 in.   30. 30.6 Ω, 42.3 Ω 31. 6.8°
32. 34.6° 33. 4.60 in.   34. 1050 ft
35. 1640 ft   36. 358,000 km   37. 243 km
38. 6.34 km   39. 9.12 in.   40. 6.35 in.

41. 1340 ft   42. 152 m   43. 9790 ft
44. 4707 ft   45. 337 ft   46. 22.9°

47. $d = \dfrac{88}{\tan A}$ 48. 429 ft, 33.3 ft

EXERCISES 16-4

1. 0.735   2. 0.482   3. 0.130   4. 0.794
5. 0.116   6. 2.01   7. 1.16   8. 0.251
9. 30.0°   10. 75.0°   11. 45.0°   12. 89.0°
13. 3.0°   14. 55.5°   15. 56.7°   16. 42.8°
17. $\dfrac{23}{41}, \dfrac{23}{41}$ 18. $\dfrac{34}{23}, \dfrac{34}{41}$ 19. $\dfrac{41}{23}, \dfrac{23}{34}$ 20. $\dfrac{34}{23}, \dfrac{41}{34}$
21. 0.549, 0.656   22. 1.52, 0.836
23. 0.549, 1.52   24. 1.82, 1.82
25. 0.385, 0.417   26. 1.67, 1.33
27. 3.78, 0.967   28. 0.845, 1.18
29. 0.461, 0.418   30. 1.09, 2.51
31. 0.898, 2.04   32. 0.736, 1.09
33. 38.0°   34. 28.4°   35. 13.1°   36. 71.0°
37. 28.4°   38. 23.2°   39. 53.4°   40. 70.0°
41. $B = 69.0°$, $a = 2.48$, $b = 6.47$
42. $A = 72.0°$, $b = 0.117$, $c = 0.379$
43. $A = 57.3°$, $a = 71.5$, $c = 85.0$
44. $A = 82.6°$, $a = 1870$, $b = 243$
45. $A = 59.1°$, $B = 30.9°$, $c = 10.1$
46. $A = 32.3°$, $B = 57.7°$, $c = 0.0142$
47. $A = 63.9°$, $B = 26.1°$, $b = 47.5$

48. $A = 71.2°$, $B = 18.8°$, $a = 51.1$
49. $B = 52.75°$, $a = 8397$, $b = 11{,}040$
50. $A = 25.667°$, $a = 2540$, $c = 5860$
51. $B = 89.98°$, $b = 12.18$, $c = 12.18$
52. $A = 0.034°$, $a = 2.116 \times 10^{-5}$, $b = 0.03566$
53. $A = 33.473°$, $B = 56.527°$, $c = 230.90$
54. $A = 14.17°$, $B = 75.83°$, $b = 0.005434$
55. $A = 88.9125°$, $B = 1.08748°$, $a = 112.483$
56. $A = 89.98°$, $B = 0.02°$, $a = 112$

57. 0.5000   58. 0.8704   59. 7.713   60. 1.054
61. 0.614 μm 62. 55.1 mA   63. $30.6°$ 64. 1210 ft·lb
65. 0.529 m   66. $42.2°$ 67. $35.9°$ 68. 77.5 ft
69. 940 ft   70. 201 m   71. $26.6°$, $63.4°$
72. 139 m   73. $28.1°$ 74. $4.9°$ 75. 0.0291 h
76. 6430 ft   77. 50 mi   78. 1030 mi   79. 575 ft
80. $16.1°$ 81. $70°$ 82. 2290 m   83. 4000 mi
84. 20.8 in. 85. 1.46 in. 86. $28.1°$
87. No; yes   88. Yes; no

EXERCISES 17-1

1. 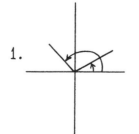 2. 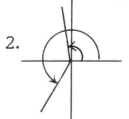 3. 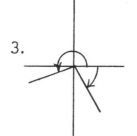 4.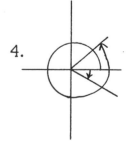

5. $\frac{12}{13}, \frac{12}{5}$ 6. $\frac{\sqrt{3}}{2}, \sqrt{3}$ 7. $-\frac{\sqrt{13}}{3}, \frac{2}{\sqrt{13}}$

8. $-\frac{15}{8}, \frac{17}{15}$ 9. $1, -\frac{1}{\sqrt{2}}$ 10. $-\frac{4}{5}, -\frac{5}{3}$

11. $\frac{3}{5}, -\frac{4}{5}$ 12. $-\frac{\sqrt{61}}{6}, \frac{5}{\sqrt{61}}$ 13. 0.316, 0.333

14. 0.461, 0.520   15. -0.413, 2.62   16. -3.37, 0.955

17. -0.772, -1.57   18. 0.167, -1.01   19. -1.85, 0.842

20. 0.227, -0.974   21. + - -   22. - + +

23. - - -   24. - - -   25. + + +

26. + + -   27. - - -   28. - - -

29. IV   30. IV   31. I

32. IV   33. II   34. IV

35. II   36. III   37. II

38. III   39. II   40. II

41. II   42. I

43. Quadrantal angle (positive y axis)   44. III

EXERCISES 17-2

1. $\sin 15°$, $-\cos 50°$ 2. $-\tan 85°$, $\sec 55°$

3. $-\cos 27°$, $-\csc 70°$ 4. $-\sec 80°$, $\cot 38°$

5. $-\tan 18°$, $\sec 10°$ 6. $-\cot 76°$, $-\sin 76°$

7. $-\cot 70°$, $\tan 60°$ 8. $\sin 40°$, $\csc 30°$

9. -0.5446   10. 0.6018   11. -10.78   12. -0.1263

13. -1.090   14. 1.054   15. -0.9265   16. -0.9291

17. 3.732   18. -2.215   19. -1.718   20. 1.026

21. 0.9759   22. -0.9888   23. 2.778   24. -1.317

25. -0.5358   26. 0.3778   27. 2.414   28. -1.235

29. 238.0°, 302.0° 30. 102.0°, 282.0°

31. 66.4°, 293.6° 32. 118.7°, 298.7°

33. 62.2°, 242.2° 34. 39.6°, 140.4°

35. 15.8°, 195.8° 36. 158.2°, 201.8°

37. 232.0°, 308.0° 38. 39.3°, 140.7°

39. 219.3°, 320.7° 40. 80.7°, 279.3°

41. 167.8°, 192.2° 42. 69.3°, 249.3°

43. 178.0°, 182.0° 44. 92.8°, 272.8°

45. 334.0° 46. 294.1° 47. 129.8° 48. 310.9°

49. 119.6° 50. 327.0° 51. 189.2° 52. 99.6°

53. 85.2° 54. 169 V   55. 354 lb   56. 41.2°

57. 334   58. 72,100 m^2 59. 2.4° 60. 32.1°

EXERCISES 17-3

1. $C = 72.6^{\circ}$, $b = 4.52$, $c = 7.23$

2. $A = 105.0^{\circ}$, $a = 68.0$, $c = 48.5$

3. $C = 109.0^{\circ}$, $a = 1390$, $b = 1300$

4. $A = 79.5^{\circ}$, $b = 1.46$, $c = 4.70$

5. $A = 149.7^{\circ}$, $C = 9.6^{\circ}$, $a = 221$

6. $A = 35.7^{\circ}$, $B = 67.9^{\circ}$, $b = 71.4$

7. $B = 8.5^{\circ}$, $C = 28.3^{\circ}$, $c = 0.733$

8. $A = 22.6^{\circ}$, $C = 44.6^{\circ}$, $c = 424$

9. $A = 99.4^{\circ}$, $b = 55.1$, $c = 24.4$

10. $C = 57.3^{\circ}$, $b = 14.6$, $c = 13.3$

11. $A = 68.1°$, $a = 552$, $c = 537$
12. $A = 90.0°$, $a = 1200$, $c = 1160$
13. $A_1 = 61.5°$, $C_1 = 70.4°$, $c_1 = 28.1$; $A_2 = 118.5°$, $C_2 = 13.4°$,
    $c_2 = 6.89$
14. $A_1 = 40.1°$, $B_1 = 124.3°$, $b_1 = 115$; $A_2 = 139.9°$, $B_2 = 24.5°$,
    $b_2 = 57.5$
15. $A_1 = 107.3°$, $C_1 = 41.3°$, $a_1 = 1060$; $A_2 = 9.9°$, $C_2 = 138.7°$,
    $a_2 = 191$
16. $B_1 = 74.5°$, $C_1 = 48.4°$, $c_1 = 0.749$; $B_2 = 105.5°$, $C_2 = 17.4°$,
    $c_2 = 0.299$
17. $B = 68.5°$, $C = 42.4°$, $b = 93.8$
18. $A = 66.9°$, $C = 11.1°$, $c = 132$
19. No solution   20. No solution
21. $B = 60.0°$, $C = 90.0°$, $b = 173$
22. $A = 18.4°$, $B = 36.4°$, $a = 9.17$
23. No solution   24. No solution
25. 15.6 in., 27.2 in.   26. 608 m, 558 m
27. 21,000 m   28. 3.59 mi
29. 19.7 km   30. 269 ft
31. 29,000 km   32. No
33. 0.618 m   34. 630 ft
35. 44.9 m   36. 1730 ft

EXERCISES 17-4

1. $A = 55.3°$, $B = 37.2°$, $c = 27.1$
2. $B = 30.7°$, $C = 107.2°$, $a = 21.4$
3. $A = 9.9°$, $C = 111.2°$, $b = 38,600$
4. $A = 20.0°$, $B = 26.1°$, $c = 14.7$
5. $A = 70.9°$, $B = 11.1°$, $c = 1580$

6. $A = 79.9°$, $C = 29.3°$, $b = 29.4$
7. $A = 18.2°$, $B = 22.2°$, $C = 139.6°$
8. $A = 65.4°$, $B = 79.7°$, $C = 34.9°$
9. $A = 42.3°$, $B = 30.3°$, $C = 107.4°$
10. $A = 47.3°$, $B = 42.1°$, $C = 90.6°$
11. $A = 51.5°$, $B = 35.1°$, $C = 93.4°$
12. $A = 42.7°$, $B = 120.2°$, $C = 17.1°$
13. $A = 46.1°$, $B = 109.2°$, $c = 138$
14. $A = 50.2°$, $B = 29.2°$, $c = 10.1$
15. $A = 132.4°$, $C = 10.3°$, $b = 4.20$
16. $B = 71.7°$, $C = 51.4°$, $a = 90.0$
17. $A = 39.9°$, $B = 56.8°$, $C = 83.4°$
18. $A = 45.5°$, $B = 78.5°$, $C = 56.0°$
19. $A = 48.6°$, $B = 102.3°$, $C = 29.1°$
20. $A = 67.5°$, $B = 59.6°$, $C = 52.9°$
21. $A = 44.37°$, $B = 60.51°$, $C = 75.12°$
22. $A = 28.23°$, $B = 42.67°$, $c = 4.315$
23. $B = 4.05°$, $C = 166.30°$, $a = 24.25$
24. $A = 57.84°$, $B = 32.71°$, $C = 89.45°$
25. 1290 m   26. 2680 mi   27. $42.8°$, $54.3°$, $82.9°$
28. $93.4°$ 29. 96.2 cm   30. 510 mi
31. $19.8°$ 32. $28.8°$ 33. $30.0°$, $60.0°$, $90.0°$
34. $29.0°$, $46.6°$, $104.5°$ 35. 158 lb
36. 13.51 cm

EXERCISES 17-5

1. $\sin \theta = \frac{3}{5}$, $\cos \theta = \frac{4}{5}$, $\tan \theta = \frac{3}{4}$, $\cot \theta = \frac{4}{3}$, $\sec \theta = \frac{5}{4}$, $\csc \theta = \frac{5}{3}$

2. $\sin \theta = \frac{12}{13}$, $\cos \theta = -\frac{5}{13}$, $\tan \theta = -\frac{12}{5}$, $\cot \theta = -\frac{5}{12}$, $\sec \theta = -\frac{13}{5}$, $\csc \theta = \frac{13}{12}$

3. $\sin \theta = -0.275$, $\cos \theta = 0.962$, $\tan \theta = -0.286$, $\cot \theta = -3.50$, $\sec \theta = 1.04$, $\csc \theta = -3.64$

4. $\sin \theta = -0.832$, $\cos \theta = -0.555$, $\tan \theta = 1.50$, $\cot \theta = 0.667$, $\sec \theta = -1.80$, $\csc \theta = -1.20$

5. II   6. III   7. IV   8. IV

9. $-\cos 48^\circ$, $\tan 14^\circ$

10. $-\sin 63^\circ$, $-\cot 42^\circ$

11. $-\sin 71^\circ$, $\sec 15^\circ$

12. $-\cos 77^\circ$, $-\csc 80^\circ$

13. -0.4540   14. 0.6561   15. -0.3057   16. 0.1405

17. -1.082   18. -9.113   19. -0.5764   20. 0.9219

21. -2.552   22. -0.3502   23. 4.230   24. -0.9914

25. 0.6820   26. -1.600   27. 1.003   28. -1.001

29. 37.0°, 217.0°

30. 212.0°, 328.0°

31. 114.9°, 245.1°

32. 151.3°, 331.3°

33. 27.4°, 152.6°

34. 45.6°, 314.4°

35. 189.4°, 350.6°

36. 74.1°, 285.9°

37. 155.0°, 335.0°

38. 8.3°, 188.3°

39. 56.3°, 123.7°

40. 123.4°, 236.6°

41. $C = 71.7^\circ$, $b = 120$, $c = 130$

42. $B = 58.5^\circ$, $b = 40.0$, $c = 45.0$

43. $A = 21.2^\circ$, $b = 128$, $c = 43.1$

44. $C = 20.0^\circ$, $a = 136$, $b = 191$
45. $A = 34.8^\circ$, $B = 53.5^\circ$, $c = 5.60$
46. $B = 56.0^\circ$, $C = 39.2^\circ$, $a = 4590$
47. $A = 59.8^\circ$, $C = 58.2^\circ$, $b = 289$
48. $B = 115.8^\circ$, $C = 6.6^\circ$, $a = 856$
49. $A_1 = 60.6^\circ$, $C_1 = 65.1^\circ$, $a_1 = 17.5$; $A_2 = 10.8^\circ$, $C_2 = 114.9^\circ$,
    $a_2 = 3.75$
50. $B = 25.9^\circ$, $C = 5.2^\circ$, $c = 158$
51. $B_1 = 68.5^\circ$, $C_1 = 60.5^\circ$, $c_1 = 73.4$; $B_2 = 111.5^\circ$, $C_2 = 17.5^\circ$,
    $c_2 = 25.3$
52. $B_1 = 33.0^\circ$, $C_1 = 127.8^\circ$, $c_1 = 2020$; $B_2 = 147.0^\circ$, $C_2 = 13.8^\circ$,
    $c_2 = 608$
53. 62.4°, 83.3°, 34.3° 54. 41.7°, 39.0°, 99.3°
55. 10.5°, 36.4°, 133.1° 56. 32.0°, 121.9°, 26.1°
57. 41.1°, 32.8°, 431 58. 0.558, 40.8°, 50.7°
59. 65.4°, 45.5°, 69.1° 60. 24.0°, 123.5°, 32.5°
61. -115 V   62. -15.0 cm
63. 585 m   64. 71.2° or 108.8°
65. 2.2° 66. 83.3°, 52.6°, 44.0° 67. 281 m
68. 14.8 m   69. 2787 ft   70. 201 km
71. 54.8° 72. 22.8°
73. 0.6788, -1.362   74. 138.2°
75. III   76. 127.2° 77. 55.064°, 58.565°, 66.370°
78. 2.373, $71^\circ46'17''$, $55^\circ36'58''$ 79. 6441.1, 5366.9, $37^\circ11'19''$
80. $B_1 = 38^\circ7'$, $C_1 = 114^\circ17'$, $c_1 = 12.31$; $B_2 = 141^\circ53'$,
    $C_2 = 10^\circ31'$, $c_2 = 2.467$

EXERCISES 18-1

1. (a) Scalar: only magnitude is specified.
   (b) Vector: magnitude and direction are specified.
2. (a) Vector: magnitude and direction are specified.
   (b) Scalar: only magnitude is specified.
3. (a) Vector: magnitude and direction are specified.
   (b) Scalar: only magnitude is specified.
4. (a) Scalar: only magnitude is specified.
   (b) Vector: magnitude and direction are specified.
5. (a) Vector: magnitude and direction are specified.
   (b) Scalar: only magnitude is specified.
6. (a) Scalar: only magnitude is specified.
   (b) Vector: magnitude and direction are specified.
7. (a) Scalar: only magnitude is specified.
   (b) Vector: magnitude and direction are specified.
8. (a) Scalar   (b) Vector

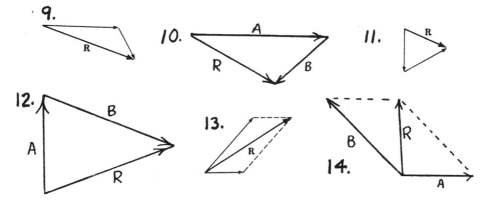

9.

10.

11.

12.

13.

14.

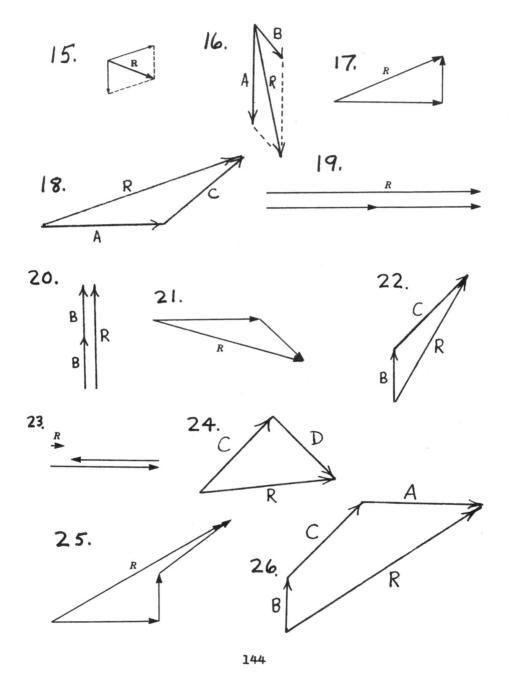

15.

16.

17.

18.

19.

20.

21.

22.

23.

24.

25.

26.

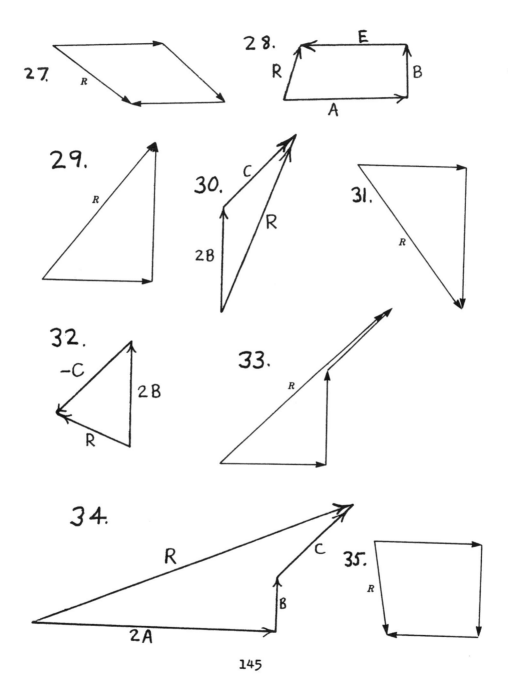

27.

28.

E

R

B

A

29.

R

30.

C

R

2B

31.

R

32.

−C

2B

R

33.

R

34.

R

C

B

2A

35.

R

145

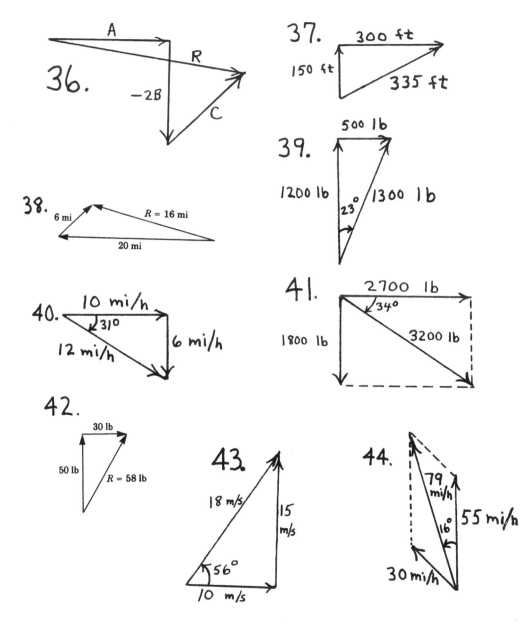

36. A   R   −2B   C

37. 300 ft   150 ft   335 ft

38. 6 mi   R = 16 mi   20 mi

39. 500 lb   1200 lb   23°   1300 lb

40. 10 mi/h   31°   12 mi/h   6 mi/h

41. 2700 lb   34°   1800 lb   3200 lb

42. 30 lb   50 lb   R = 58 lb

43. 18 m/s   15 m/s   56°   10 m/s

44. 79 mi/h   16°   55 mi/h   30 mi/h

146

EXERCISES 18-2

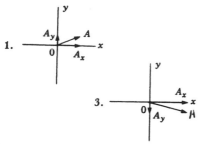

5. $A_x = 25.4$, $A_y = 12.9$

6. $A_x = 122$, $A_y = 457$

7. $A_x = -8.60$, $A_y = 57.6$

8. $A_x = -5.98$, $A_y = 4.88$

9. $A_x = 3340$, $A_y = -5590$

10. $A_x = 86.6$, $A_y = -40.4$

11. $A_x = -701$, $A_y = -229$

12. $A_x = -0.502$, $A_y = -0.641$

13. $A_x = -1344$, $A_y = -3212$

14. $A_x = -1049$, $A_y = -7968$

15. $A_x = -0.1306$, $A_y = 0.8636$

16. $A_x = -0.5160$, $A_y = 0.1607$

17. $A_x = 18.2$, $A_y = 18.2$

18. $A_x = 403$, $A_y = 233$

19. $A_x = 3.45$, $A_y = 5.32$

20. $A_x = 129$, $A_y = 42.7$

21. $A_x = -26.7$, $A_y = -217$

22. $A_x = -32.0$, $A_y = -12.9$

23. $A_x = -969$, $A_y = -969$

24. $A_x = 20.7$, $A_y = -20.7$

25. $A_x = 52.4$, $A_y = -69.7$

26. $A_x = 208$, $A_y = -67.7$

27. $A_x = -0.03233$, $A_y = -0.04370$

28. $A_x = -0.00945$, $A_y = 0.0181$

29. $A_x = 25815$, $A_y = -1714.6$

30. $A_x = 2359.6$, $A_y = -13520$

31. $A_x = -13.378$, $A_y = 31.981$

32. $A_x = -56.672$, $A_y = 13.710$

33. 33.3 mi (east); 14.8 mi (north)
34. 30 mi/h (north); 30 mi/h (west)
35. 211 N (left); 751 N (vertical)
36. 8080 lb   37. 0.013 (x axis); 0.020 (y axis)
38. 14,000 lb   39. 3.72 A, 39.1° 40. 1160 lb

EXERCISES 18-3

1. $R = 5.39$, $\theta = 21.8^{\circ}$ 2. $R = 73.5$, $\theta = 11.8^{\circ}$
3. $R = 1460$, $\theta = 59.2^{\circ}$ 4. $R = 1.04$, $\theta = 21.8^{\circ}$
5. $R = 38.3$, $\theta = 25.2^{\circ}$ 6. $R = 714$, $\theta = 79.8^{\circ}$
7. $R = 25.49$, $\theta = 61.8^{\circ}$ 8. $R = 0.0631$, $\theta = 56.5^{\circ}$
9. $R = 10.7$, $\theta = 23.7^{\circ}$ 10. $R = 750$, $\theta = 338.9^{\circ}$
11. $R = 276$, $\theta = 55.3^{\circ}$ 12. $R = 7.44$, $\theta = 331.8^{\circ}$
13. $R = 115$, $\theta = 102.0^{\circ}$ 14. $R = 0.510$, $\theta = 37.4^{\circ}$
15. $R = 121$, $\theta = 272.1^{\circ}$ 16. $R = 22.3$, $\theta = 235.2^{\circ}$
17. $R = 10$, $\theta = 36.9^{\circ}$ 18. $R = 1.16$, $\theta = 61.6^{\circ}$
19. $R = 61.1$, $\theta = 116.8^{\circ}$ 20. $R = 713$, $\theta = 38.5^{\circ}$
21. $R = 1560$, $\theta = 201.6^{\circ}$ 22. $R = 24.8$, $\theta = 324.3^{\circ}$
23. $R = 7675.1$, $\theta = 175.07^{\circ}$ 24. $R = 0.009888$, $\theta = 214.7^{\circ}$
25. $R = 13$, $\theta = 293^{\circ}$ 26. $R = 179$, $\theta = 300.7^{\circ}$
27. $R = 28.9$, $\theta = 327.5^{\circ}$ 28. $R = 844$, $\theta = 320.5^{\circ}$
29. $R = 4352$, $\theta = 321.0^{\circ}$ 30. $R = 0.392$, $\theta = 138.8^{\circ}$
31. $R = 0.321$, $\theta = 193.7^{\circ}$ 32. $R = 20.20$, $\theta = 320.9^{\circ}$
33. 264 lb at an angle that is 36.7° from the 212 lb force.
34. 14.0 lb at an angle of 27.7° below the horizontal.
35. 46.0 mi, 34.4° S of E

36. 7.70 km/h, 24.6° away from the line straight across

37. 3990 km/h, 5.0° from direction of plane

38. 26.2 lb, 38.3° from the 58.0 lb force

39. 53.9 lb, $\theta = 68.2°$

40. 605 mi/h, 85.0° north of west

41. 130 V at an angle of 6.9° from V_1

42. 163 A/m at an angle of 22.1° from H_1

43. Yes, to the right.   44. Yes, since 19.1 lb is greater than 18.0 lb.

EXERCISES 18-4

1. (a) Scalar: only magnitude is specified.

   (b) Vector: magnitude and direction are specified.

2. (a) Vector: magnitude and direction are specified.

   (b) Scalar: only magnitude is specified.

3. (a) Scalar: only magnitude is specified.

   (b) Vector: magnitude and direction are specified.

4. (a) Vector: magnitude and direction are specified.

   (b) Scalar: only magnitude is specified.

6.

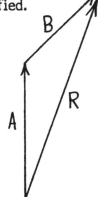

5.

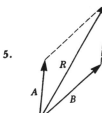

7.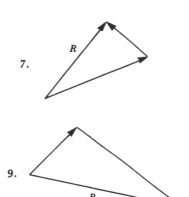

8.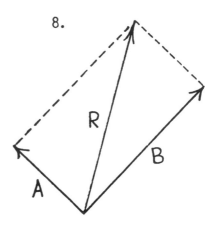

9.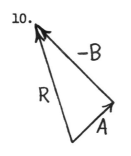

10.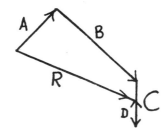

11.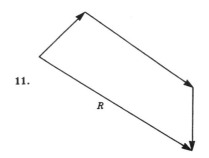

12.

13.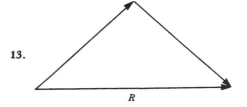

14.

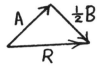

15. R

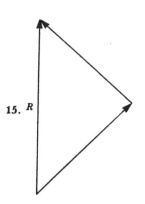

16.

$2D$ $-C$

$\uparrow R$

17. $A_x = 6.77$, $A_y = 2.73$ 18. $A_x = 15.1$, $A_y = 44.0$

19. $A_x = -1380$, $A_y = 778$ 20. $A_x = 0.448$, $A_y = -0.788$

21. $A_x = 15.9$, $A_y = 6.50$ 22. $A_x = -85.6$, $A_y = 93.8$

23. $A_x = 8.903$, $A_y = -13.87$ 24. $A_x = -0.111$, $A_y = -0.890$

25. 35 mi/h (west), 9.3 mi/h north

26. 254 mi/h (west), 92.3 mi/h (south)

27. 431 lb (down), 349 lb (horizontally)

28. 2.4 lb (horizontally), 27 lb (vertically)

29. $R = 38.8$, $\theta = 27.7°$ 30. $R = 1060$, $\theta = 62.4°$

31. $R = 5530$, $\theta = 81.7°$ 32. $R = 2190$, $\theta = 16.1°$

33. $R = 27.66$, $\theta = 40.0°$ 34. $R = 753$, $\theta = 23.4°$

35. $R = 1.10$, $\theta = 48.0°$ 36. $R = 52.1$, $\theta = 125.5°$

37. $R = 47$, $\theta = 314°$ 38. $R = 7.18$, $\theta = 74.8°$

39. $R = 64.8$, $\theta = 68.7°$ 40. $R = 656$, $\theta = 269.6°$

41. 113 lb   42. $A_x = -4510$ lb, $A_y = -32100$ lb

43. 348 km/h (horizontally), 2220 km/h (vertically)

44. $A_x = 367$ ft/s (horizontally), $A_y = 711$ ft/s (vertically)

45. 232 lb

46. $A_x = 353$ mi/h (horizontally), 140 mi/h (vertically)

47. 286 mi   48. $R = 1470$ lb, $\theta = 54.7°$ below the horizontal force

49. $R = 13,500$ lb, $\theta = 45.8°$ (above the horizontal)

50. $R = 88.3$ lb, $\theta = 8.93°$ from the 87.2 lb force

51. $R = 1180$ lb, $\theta = 43.0°$ (away from the force of 532 lb)

52. 32 lb   53. 374 lb

54. $R = 356$ N, $\theta = 29.6°$ from the 176 N force

55. 61.0 mi, $35.0°$ N of E   56. 5.66 km, $8.68°$ N of W

57. 153 V   58. 2650 mi   59. $R = 868.6$, $\theta = 6.8°$ N of E

60. 102 m, $37.1°$ S of E   61. $R = 24$ m, $\theta = 36.5°$ N of E

62. 118 km, $12.2°$ N of E   63. 15.3 mi/h, $11.3°$ off ship's path

64. 9.92 m/s, $40.9°$ (above horizontal)   65. 167 mi/h, $84.2°$ S of E

66. 29 km, $2.9°$ S of E   67. 30 lb, $53.4°$ from the 32 lb force

68. 7100 lb, $47.3°$ from the side force

EXERCISES 19-1

1. $\frac{2\pi}{9}, \frac{4\pi}{45}$ 2. $\frac{\pi}{5}, \frac{4\pi}{3}$ 3. $\frac{11\pi}{36}, \frac{11\pi}{6}$ 4. $\frac{\pi}{9}, \frac{67\pi}{36}$

5. $\frac{\pi}{6}, \frac{3\pi}{4}$ 6. $\frac{\pi}{3}, \frac{5\pi}{3}$ 7. $\frac{35\pi}{36}, \frac{7\pi}{6}$ 8. $\frac{13\pi}{45}, \frac{14\pi}{9}$

9. $120°, 36°$ 10. $18°, 144°$ 11. $15°, 225°$
12. $40°, 270°$ 13. $168°, 140°$ 14. $165°, 288°$
15. $35°, 135°$ 16. $306°, 345°$ 17. 0.802
18. 1.26   19. 3.31   20. 4.41
21. 4.86   22. 1.72   23. 3.18
24. 5.72   25. $46°$ 26. $14°$
27. $143°$ 28. $100°$ 29. $186°$
30. $286°$ 31. $710°$ 32. $4297°$
33. 0.5000   34. 0.7071   35. -1.732
36. -0.2588   37. 0.8674   38. 0.7074
39. 1.197   40. 0.1417   41. 1.556
42. 1.082   43. 7.086   44. 1.231
45. 0.85730   46. 0.1460   47. -1.145
48. -1.001   49. -9.657   50. 0.3312
51. 3.796   52. -1.000   53. 1.047, 5.236
54. 1.500, 1.642   55. 1.9116, 5.0532   56. 2.482, 3.802
57. 0.4500, 2.692   58. 0.2500, 3.392   59. 4.470, 4.954
60. 1.180, 5.103   61. $0.147°$ 62. 52 m/s
63. 160 V   64. Second pulley $(343.8°)$
65. 33.4 m   66. 0.1 s   67. $1620°$, 9π rad
68. $828,000°$; 4600π rad   69. 0.98339
70. 7.4448   71. 0.66300   72. 0.28143

153

73. 0.4814
74. 5.712
75. 0.151504

76. 0.113665
77. 146°
78. 78.78°

79. 13.8°
80. 143°
81. -0.315209

82. 0.80424
83. 0.000421
84. 0.000786

EXERCISES 19-2

1. (a) 7.34 cm; (b) 13.4 cm²
2. (a) 8.262 in.; (b) 65.19 in.²

3. (a) 355 mm; (b) 73,000 mm²
4. (a) 0.473 m; (b) 0.0651 m²

5. (a) 0.620 ft; (b) 0.735 ft²
6. (a) 19.4 cm; (b) 90.0 cm²

7. (a) 26.9 in.; (b) 88.0 in.²
8. (a) 9420 mi; (b) 18,700,000 mi²

9. 70.6 cm/s
10. 15,350 mm/min
11. 6920 ft/min

12. 10,700 in./s
13. 37.7 cm
14. 3.90 m

15. 679 cm²
16. 510.6 m²
17. 1.62

18. 54.0° or 0.942
19. 87.8 ft²
20. 4.80 ft²

21. 0.114 m
22. 69.1 mi
23. 2.0 mi

24. 15.1 cm
25. 3130 cm
26. 391 cm

27. 26.4 ft
28. 0.0431 mi
29. 59 rad/s

30. 27,200 mi/h
31. 2.6×10^{-6} rad/s
32. 2710 m/min

33. 153,000 cm/min
34. 20.6 r/min
35. 13,200 ft/min

36. 40,600 ft/min
37. 134 ft²
38. 148 cm²

39. 4.959 in./s
40. 85.9 ft, 86.7 ft
41. 4.85×10^{-6}

42. 0.000009
43. 80,400,000 mi
44. 0.419 ft

EXERCISES 19-3

1.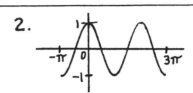

0, −0.7, −1, −0.7, 0, 0.7, 1, 0.7, 0,

−0.7, −1, −0.7, 0, 0.7, 1, 0.7, 0

2.

−1, −0.7, 0, 0.7, 1, 0.7,
0, −0.7, −1, −0.7, 0, 0.7,
1, 0.7, 0, −0.7, −1

3.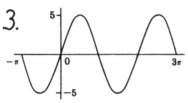

0, −3.5, −5, −3.5, 0, 3.5, 5, 3.5, 0,

−3.5, −5, −3.5, 0, 3.5, 5, 3.5, 0

4.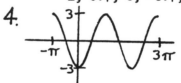

3, 2.1, 0, −2.1, −3, −2.1,
0, 2.1, 3, 2.1, 0, −2.1,
−3, −2.1, 0, 2.1, 3

5. $y = 3 \sin x$

6. $y = 5 \cos x$

7. $y = -6 \sin x$

8. $y = -\dfrac{1}{2} \cos x$

9.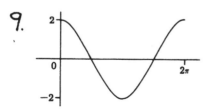

10.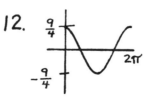

11.

12.

155

13.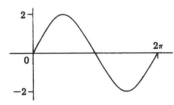

14.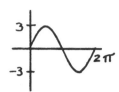

15.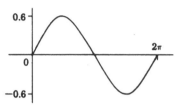

16.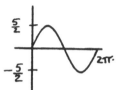

17.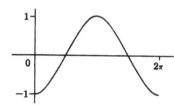

18.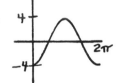

19.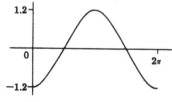

20.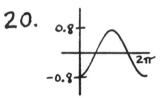

21.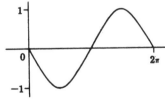

22.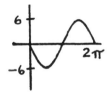

23.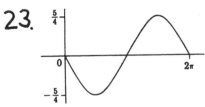

24.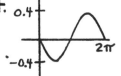

156

25.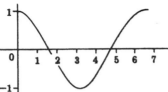

26.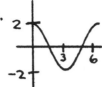

27.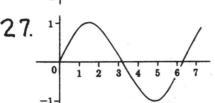

28.

29.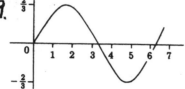

30.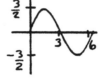

31.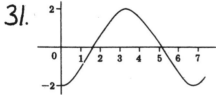

32.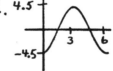

33.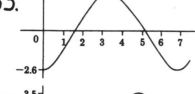

34.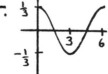

35.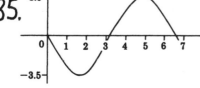

36.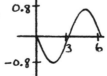

157

EXERCISES 19-4

1. $\frac{2\pi}{3}$ 2. $\frac{2\pi}{3}$ 3. $\frac{\pi}{2}$ 4. $\frac{\pi}{5}$

5. $\frac{2\pi}{3}$ 6. $\frac{\pi}{5}$ 7. $\frac{\pi}{3}$ 8. $\frac{\pi}{4}$

9. $\frac{2\pi}{5}$ 10. $\frac{\pi}{3}$ 11. 1 12. $\frac{1}{5}$

13. 4π 14. 6π 15. 3π 16. 8π

17. $\frac{4}{3}$ 18. 6 19. $\frac{2}{\pi}$ 20. $2\pi^2$

21. 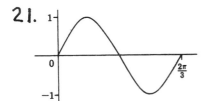   22.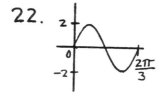

23. 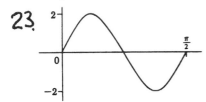   24.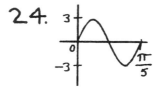

25. 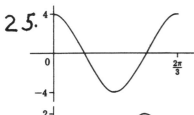   26.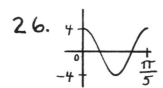

27. 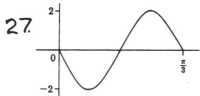   28.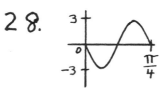

29.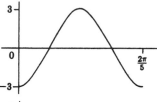

30.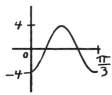

31.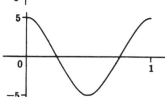

32.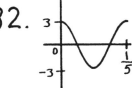

33.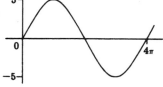

34.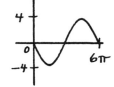

35.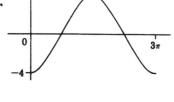

36.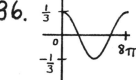

37.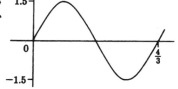

38.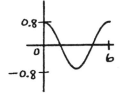

39.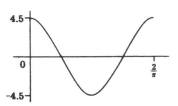

40.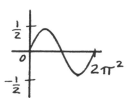

41. $y = \sin 2x$

42. $y = \sin 6x$

43. $y = \sin \pi x$

44. $y = \sin \frac{2\pi}{3}x$

45. $y = \sin 2\pi x$

46. $y = \sin \frac{\pi}{3}x$

47. $y = \sin 5x$

48. $y = \sin \frac{14}{3}x$

49. $y = 2 \sin \pi x$

50. $y = 3 \cos \pi x$

51. $y = 5 \sin \frac{\pi}{2}x$

52. $y = -5 \cos \frac{\pi}{4}x$

53.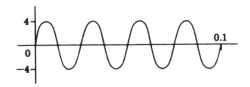

54.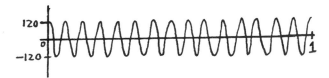

55.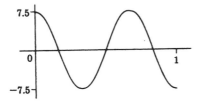

56.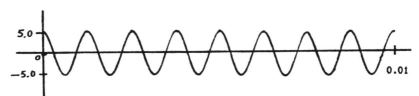

57.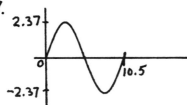

58.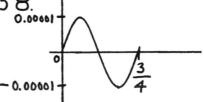

59.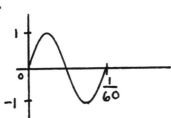

60.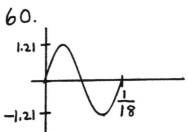

EXERCISES 19-5

1. $1, 2\pi, -\dfrac{\pi}{3}$

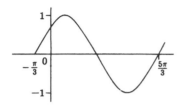

2. $2, 2\pi, \dfrac{\pi}{3},$

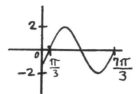

3. $1, 2\pi, \dfrac{\pi}{3}$

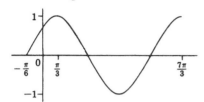

4. $5, 2\pi, -\dfrac{\pi}{4}$

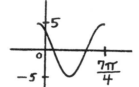

5. $3, \pi, -\dfrac{\pi}{8}$

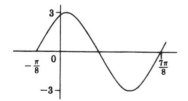

6. $1, \pi, \dfrac{\pi}{8}$

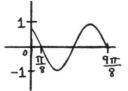

7. $4, \dfrac{2\pi}{3}, -\dfrac{\pi}{3}$

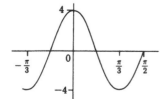

8. $6, \pi, \dfrac{\pi}{4}$

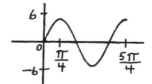

162

9. $2, 6\pi, -\dfrac{3\pi}{2}$

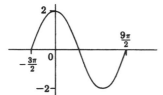

10. $\dfrac{1}{5}, 8\pi, 2\pi$

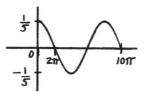

11. $6, 3\pi, -\dfrac{\pi}{2}$

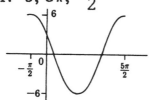

12. $\dfrac{1}{6}, 4\pi, 2\pi$

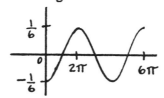

13. $10, 2, \dfrac{1}{\pi}$

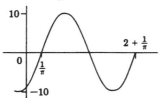

14. $8, 1, -\dfrac{1}{2}$

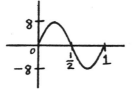

15. $\dfrac{3}{2}, \dfrac{2}{3}, \dfrac{1}{9\pi}$

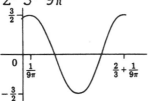

16. $\dfrac{3}{5}, \dfrac{1}{2}, -\dfrac{1}{20}$

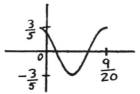

17. $1.2, 2, \dfrac{1}{6}$

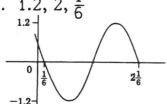

18. $0.4, 1, -\dfrac{1}{8\pi}$

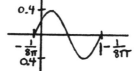

19. $6, 2, \frac{1}{2\pi}$

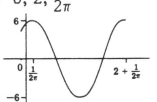

20. $8, \frac{2}{5}, -\frac{1}{50\pi}$

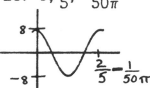

21. $\frac{3}{4}, \frac{2}{\pi}, -\frac{1}{\pi}$

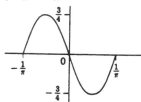

22. $1, \frac{\pi}{2}, \frac{1}{2\pi}$

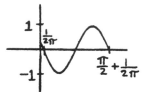

23. $\frac{5}{2}, 2, \frac{\pi}{3}$

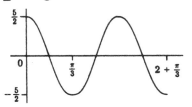

24. $\pi, 2\pi^2, \frac{3\pi}{2}$

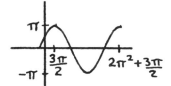

25.

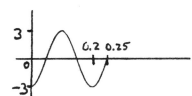

26.

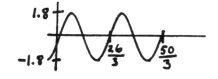

27.

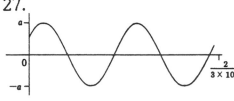

28.

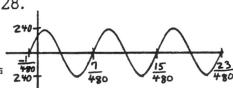

164

29.

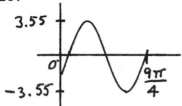

30.

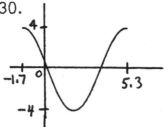

31.

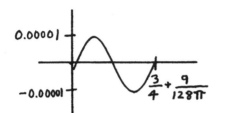

32.

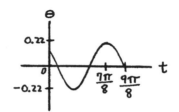

EXERCISES 19-6

1. $\dfrac{7\pi}{20}, \dfrac{137\pi}{180}$

2. $\dfrac{3\pi}{8}, \dfrac{9\pi}{8}$

3. $\dfrac{23\pi}{90}, \dfrac{\pi}{18}$

4. $\dfrac{53\pi}{30}, \dfrac{71\pi}{45}$

5. $20°, 150°$

6. $216°, 112.5°$

7. $70°, 81°$

8. $342°, 252°$

9. $35.8°$

10. $76.2°$

11. $197.7°$

12. $709.3°$

13. 1.31

14. 1.920

15. 5.934

16. 0.271

17. 2.662

18. 3.759

19. 0.16

20. 7.365

21. 0.6894, 2.452

22. 0.1226, 6.161

23. 1.998, 4.285

24. 1.124, 4.266

25. (a) 7.73 cm; (b) 47.9 cm²   26. (a) 846 ft; (b) 214,000 ft²
27. (a) 2.27 in.; (b) 8.23 in.²   28. (a) 2.86 m; (b) 0.977 m²
29. 4.60 m/min   30. 2270 in./min
31. 353 m/min   32. 187,000 cm/min

33. 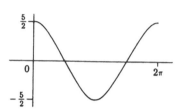   34.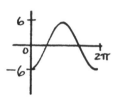

35. 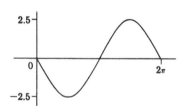   36.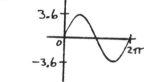

37. 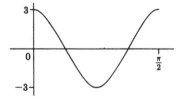   38.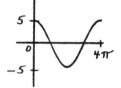

39. 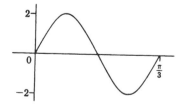   40.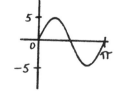

41.

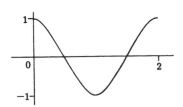

42.

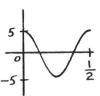

43.

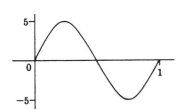

44.

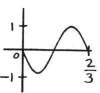

45.

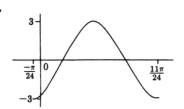

46.

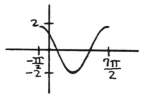

47.

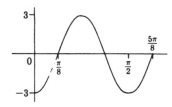

48.

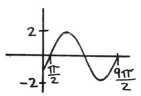

49.

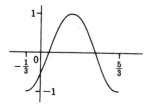

50.

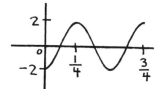

51.

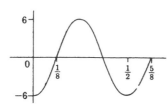

52.

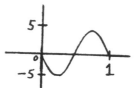

53. $y = 8 \sin \pi x$ 54. $y = 4 \cos \pi x$ 55. $y = -3 \cos 2\pi x$

56. $y = -5 \sin \frac{\pi}{4}x$ 57. 8.78 cm   58. 89.8 mA

59. 2670 mi   60. 9.50 cm   61. 117.0° or 2.04 rad

62. 5900 mi/h   63. 36.3 cm²   64. 864 cm²

65. 126,000 cm/min   66. 4900 ft/min

67. 138 rad/s or 21.9 r/s   68. 0.153 m

69. 905 r/min   70. 112,000 in./min

71. 1040 mi/h   72. 67.8 mi   74. 0.051 m   75. 21 V

73.

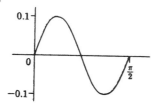

76.

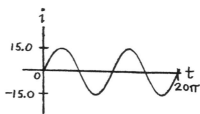

77.

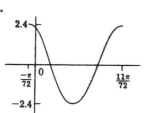

78.

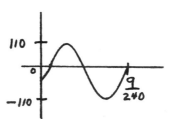

79. 374,000 km   80. 7560 mi

EXERCISES 20-1

1. $4j$ 2. $5j$ 3. $-3j$ 4. $-6j$

5. $0.5j$ 6. $-0.4j$ 7. $3j\sqrt{3}$ 8. $3j\sqrt{5}$

9. $-4j\sqrt{3}$ 10. $-5j\sqrt{3}$ 11. $0.02j$ 12. $-0.03j$

13. $5 + j$ 14. $-3 + j$ 15. $6 - 2j$ 16. $-9 - 3j$

17. $-4 + 2j\sqrt{2}$ 18. $-8 - 3j\sqrt{6}$ 19. $14 - 3j\sqrt{7}$

20. $-7 + 2j\sqrt{7}$ 21. -9 22. -4

23. -6 24. -0.5 25. 3 26. 7

27. $-2\sqrt{3}$ 28. -4 29. -6 30. $-\sqrt{35}$

31. $-\sqrt{33}$ 32. $-\sqrt{65}$ 33. -1 34. 1

35. j 36. -1 37. j 38. $-j$

39. -1 40. 1 41. -1 42. j

43. $-j$ 44. 1 45. $3 + 8j$ 46. 7

47. -4 48. $-5 - 4j$ 49. $-9j$ 50. $3j$

51. $-2 + 7j$ 52. $3 + 8j$

EXERCISES 20-2

1. $5 - 7j$ 2. $8 - j$ 3. $4 + 5j$

4. $-12 + 3j$ 5. $1 + 3j$ 6. $8 - 9j$

7. $-12j$ 8. $2 - 15j$ 9. $2 - 6j$

10. $10 + j$ 11. $-3 + 20j$ 12. $11 - 3j$

13. -30 14. 6 15. 5

16. 25 17. $48 - 9j$ 18. $-66 + 42j$

19. $36j$ 20. $-24j$ 21. $-6 + 4j$

22. $-12 - 9j$ 23. $36 + 3j$ 24. $-6 - 9j$

25. $\dfrac{-8 + 6j}{25}$ 26. $\dfrac{-15 + 6j}{29}$ 27. $\dfrac{8 - 3j}{2}$

28. $\dfrac{6 + 5j}{3}$

29. $\dfrac{53 - 27j}{61}$

30. $\dfrac{11 - 3j}{15}$

31. $\dfrac{-5 - j}{2}$

32. $\dfrac{-23 - 24j}{13}$

33. $\dfrac{3 + 3j}{4}$

34. $\dfrac{-3 + 4j}{5}$

35. $\dfrac{-3 + 7j}{6}$

36. $\dfrac{31 - 29j}{34}$

37. $-2j$

38. $-5 - 12j$

39. $2 - 3j$

40. $-9 + 6j$

41. $-30 - 40j$

42. $-84 + 135j$

43. $3 + 4j$

44. $\dfrac{-89 - 140j}{29}$

45. $-10 + 6j$

46. $39 + 24j$

47. $-48j$

48. $-28 - 44j$

49. 100

50. 34

51. 53

52. 10

53. $14.7 - 2.1j$ ohms

54. $\dfrac{15 + 7j}{4}$ ohms

55. $3.1 + 5.8j$ volts

56. $\dfrac{89 + 23j}{26}$ ohms

EXERCISES 20-3

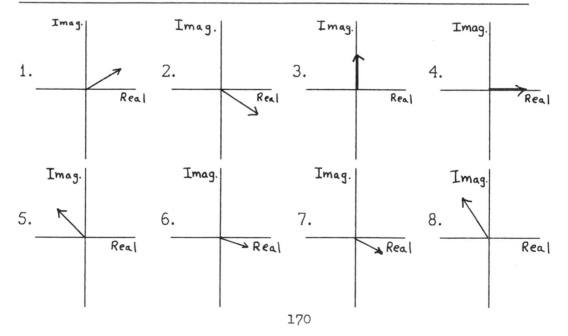

170

9. $5 + 4j$ 10. $5 + 10j$ 11. $2 + 6j$

12. $3 + 3j$ 13. $6 + j$ 14. $8 - j$

15. $7 + j$ 16. -8 17. $4 - j$

18. $-3 - 4j$ 19. $9 - 15j$ 20. $5 + 7j$

21. 4 22. $9 + 2j$ 23. $-13 + 2j$

24. $8j$ 25. $-15 - 11j$ 26. $3 + 13j$

27. $12 - 4j$ 28. $3 + 5j$

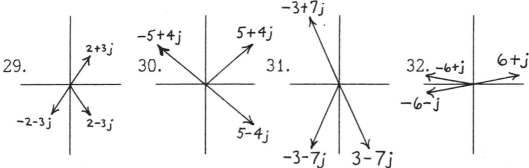

EXERCISES 20-4

1. $5(\cos 37° + j \sin 37°)$ 2. $5(\cos 143° + j \sin 143°)$

3. $5(\cos 323° + j \sin 323°)$ 4. $5(\cos 217° + j \sin 217°)$

5. $4.50(\cos 302.9° + j \sin 302.9°)$

6. $4.46(\cos 159.5° + j \sin 159.5°)$

7. $6.52(\cos 55.8° + j \sin 55.8°)$

8. $1060(\cos 216.5° + j \sin 216.5°)$

9. $2(\cos 225° + j \sin 225°)$ 10. $2(\cos 330° + j \sin 330°)$

11. $2(\cos 120° + j \sin 120°)$ 12. $2(\cos 60° + j \sin 60°)$

13. $10(\cos 180° + j \sin 180°)$ 14. $7(\cos 0° + j \sin 0°)$

15. $6(\cos 90° + j \sin 90°)$ 16. $3(\cos 270° + j \sin 270°)$

17. $6.15(\cos 43.0° + j \sin 43.0°)$

18. $256(\cos 184.6^\circ + j \sin 184.6^\circ)$
19. $10.3(\cos 335.2^\circ + j \sin 335.2^\circ)$
20. $73.5(\cos 109.6^\circ + j \sin 109.6^\circ)$
21. $0.348(\cos 76.4^\circ + j \sin 76.4^\circ)$
22. $8.19(\cos 27.4^\circ + j \sin 27.4^\circ)$
23. $56.0(\cos 212.5^\circ + j \sin 212.5^\circ)$ 24. $403(\cos 8.3^\circ + j \sin 8.3^\circ)$
25. $1.03 + 2.82j$ 26. $-2.11 - 4.53j$ 27. $1.17 - 0.940j$
28. $-1.11 - 4.36j$ 29. $-10j$ 30. 20
31. -25   32. $65j$ 33. $41 + 29j$
34. $-38 - 65j$ 35. $6.2 - 11j$ 36. $-2.54 - 1.47j$
37. $10.06 + 7.380j$ 38. $-728 + 213j$ 39. $-1.258 - 1.797j$
40. $-3.171 + 4.765j$ 41. $56.8 + 7.18j$ volts
42. $3.95 + 1.56j$ microamperes
43. Magnitude: 81.2 lb; direction: 322.9°
44. Magnitude: 0.455 mm; direction: 20.3°

EXERCISES 20-5

1. $8j$ 2. $-10j$ 3. $-20j$
4. $9j$ 5. $3j\sqrt{6}$ 6. $2j\sqrt{10}$
7. $-2j\sqrt{14}$ 8. $-3j\sqrt{7}$ 9. $-6 + 10j$
10. $5 - 12j$ 11. $3 - 4j\sqrt{3}$ 12. $-2 + 3j\sqrt{3}$
13. -25   14. 8   15. $-\sqrt{14}$
16. -8   17. -1   18. $-j$
19. 1   20. j 21. $14 + 4j$
22. $1 - 11j$ 23. $-1 - 7j$ 24. $-15 + j$
25. $15j$ 26. $32 + 4j$ 27. $-11 + 27j$

28. $103 + 11j$ 29. $\dfrac{-2 - 16j}{5}$ 30. $\dfrac{-21 - j}{17}$

31. $\dfrac{-46 - 2j}{53}$ 32. $-3 + 2j$

33. Sum: 12; product: 100   34. Sum: -4; product 29

35. Sum: 14; product: 58   36. Sum: -8; product: 65

37. $3 + 3j$ 38. $-5 + 7j$

39. $-9 - 7j$ 40. $-6 + 15j$

41. $13(\cos 67^\circ + j \sin 67^\circ)$ 42. $8.47(\cos 330.6^\circ + j \sin 330.6^\circ)$

43. $21.8(\cos 214.3^\circ + j \sin 214.3^\circ)$

44. $107(\cos 125.6^\circ + j \sin 125.6^\circ)$

45. $5.15 + 14.3j$ 46. $-0.734 - 0.541j$

47. $14.1 - 8.97j$ 48. $-3.40 + 0.593j$

49. Yes; yes

50. $(a + bj)(a - bj) = a^2 - b^2 j^2 = a^2 + b^2$

51. $9 + 42j$ volts   52. $\dfrac{46 - 140j}{89}$ amperes

Exercises from Appendix C

1. 3   2. 6   3. 9   4. 22   5. 108
6. 67   7. 305   8. 451   9. 110   10. 1010
11. 10001   12. 10011   13. 101110   14. 110011   15. 1001111
16. 10010001 17. 101   18. 1000   19. 1111   20. 11000
21. 101101   22. 110100 23. 11011001   24. 10101001
25. 110   26. 1010   27. 1111   28. 10010
29. 1110101   30. 10001111   31. 100011011010
32. 100000111010   33. 110   34. 100   35. 11101
36. 11110   37. 11010   38. 1101   39. 1011110   40. 1000011